1 MONTH OF
FREE
READING

at

www.ForgottenBooks.com

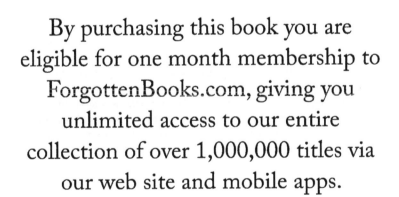

By purchasing this book you are eligible for one month membership to ForgottenBooks.com, giving you unlimited access to our entire collection of over 1,000,000 titles via our web site and mobile apps.

To claim your free month visit:
www.forgottenbooks.com/free903347

ISBN 978-0-265-87804-0
PIBN 10903347

UNITED STATES
DEPARTMENT OF AGRICULTURE

CIRCULAR No. 544

Washington, D. C.  April 1940

METHODS OF VENTILATING WHEAT IN FARM STORAGES

By

C. F. KELLY

Associate Agricultural Engineer
Division of Farm Structures Research
Bureau of Agricultural Chemistry and Engineering

For sale by the Superintendent of Documents, Washington, D. C. · · · · · · · Price 15 cents

CIRCULAR No. 544   APRIL 1940

UNITED STATES DEPARTMENT OF AGRICULTURE

WASHINGTON, D. C.

METHODS OF VENTILATING WHEAT IN FARM STORAGES

By C. F. Kelly, *associate agricultural engineer, Division of Farm Structures Research, Bureau of Agricultural Chemistry and Engineering* [1]

CONTENTS

INTRODUCTION

A considerable part of every wheat crop is stored on farms before it goes to market. Improved and expanded farm storage facilities would, in many cases, aid growers in marketing their wheat to advantage. The best practice is to thresh or combine wheat only when it is dry enough to store safely; but wet weather at harvest or threshing time, too early use of the combine, or green weed seeds in the threshed grain may cause wheat to have too high an initial moisture content for safe storage on the farm.

Under such conditions, simple and reliable methods of holding and drying damp grain without spoilage would be valuable. If inexpensive and effective methods for ventilating grain in the bin were available, the problem would be partly solved. Farmers have felt the need for and have experimented with bin ventilation for years, but no method has as yet been devised that is universally satisfactory.

This circular presents the record and findings of a series of experiments on the effect of bin ventilation on the storage and conditioning of damp wheat in the laboratory and in full-sized bins in Kansas,

[1] The author acknowledges the assistance of S. J. Dennis, associate refrigerating engineer and B. M. Stahl, associate agricultural engineer, in preparing this circular.

159482°—40——1

Illinois, Maryland, and North Dakota, during the seasons of 1936, 1937, and 1938. These studies are part of an investigation of grain storage on the farm which was conducted under authorization of the Secretary of Agriculture with funds provided under the Bankhead-Jones Act, approved June 29, 1935. In the work reported in this circular the United States Department of Agriculture cooperated with the agricultural experiment stations of the States where the work was conducted.[2]

The Department was represented by the Bureau of Agricultural Engineering which handled the construction of the bins, storage of

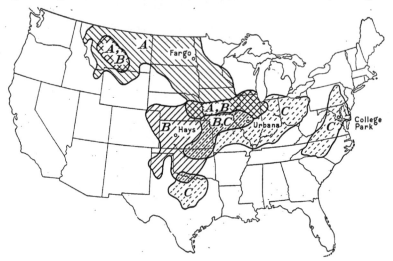

FIGURE 1.—Principal wheat-producing areas and the locations at which experimental work was conducted: *A*, Hard red spring wheat area; *B*, hard red winter wheat area; *C*, soft red winter wheat area.

wheat, and reading of temperatures and assisted in taking of samples; the Bureau of Agricultural Economics, which sampled and graded the wheat according to the United States official grain standards and made fat-acidity, milling, and baking tests; and by the Bureau of Plant Industry, which made germination tests and studies of mold growth.

The general objectives of this investigation were to determine what types of storage structure will best preserve and improve the quality of wheat stored on farms, and to find the maximum moisture contents of the various classes of wheat for safe storage in farm structures of various designs.

The objects of the portion of the work reported in this circular were to find the most practical methods for checking spoilage and aiding the drying of damp wheat in farm storage.

In order to learn the influence of differences in climate, harvesting practices, and classes of wheat upon the requirements for storage, experiments were carried on at the four widely separated locations shown in figure 1. College Park, Md., and Urbana, Ill., represent the soft red winter wheat area, with harvest coming in late June and early July. Hays, Kans., is in the hard red winter wheat area with

[2] The State Mill and Elevator of North Dakota loaned wheat for experimental use in 1936 and 1937.

harvest late in June, and Fargo, N. Dak., is in the hard red spring wheat area with harvest in August. In Maryland, wheat for farm storage is usually cut with the binder and left in the shock or stack until dry enough to be threshed. In Illinois and North Dakota both shock threshing and combining are common, while in Kansas combining is becoming the prevailing practice.

The field studies of bin ventilation were made in bins of 20 to 1,000 bushels capacity. These were filled with moderately damp wheat, then closely observed for as long as was necessary to determine the value of the equipment and methods used. The different ventilating systems were compared by noting the temperature changes of the grain in different parts of the bin, the rates at which moisture was removed from the stored wheat, the commercial grade of the wheat when placed in storage and during the storage period, and the changes in the condition of the wheat as indicated by its fat acidity and percent germination.

Temperatures of the stored wheat during the storage period were obtained by thermocouples placed in representative locations and read by means of a potentiometer. Average samples were taken from the bins with standard grain probes and analyzed by skilled workers to determine moisture content, grade, germination, and fat acidity. In order to learn the condition of the wheat in different parts in the bins, additional samples were taken from selected locations and tested for moisture content, fat acidity, and germination.

The amount of grain in the experimental bins was estimated on a volume basis, assuming 1¼ cubic feet equals 1 bushel (60 pounds) of wheat.

FUNDAMENTALS OF WHEAT STORAGE

The first step in this investigation was to collect and organize information bearing on the physical properties of wheat, and its reaction to different environmental conditions. Much information of this kind was available from the work of other investigators, as noted in subsequent references; new laboratory studies were made as reported in the appendix. The following is a brief summary of basic information related to this problem.

CAUSES AND NATURE OF DAMAGE IN STORED WHEAT

Rapid deterioration of damp wheat in storage is usually caused by the activity of micro-organisms and by excessive respiration of the wheat kernel. Heat is evolved both by ever-present micro-organisms and by kernel respiration at a rate depending to a large extent on the wheat moisture content and wheat temperature. Where the heat generated by excessive respiration is not removed promptly, it will raise the wheat temperature and increase the rate of respiration. Unless the heat and excess moisture are removed, this self-accelerating process will continue until the grain is unfit for human consumption.

The maximum safe moisture content for long-time storage in non-ventilated farm structures varies in the different wheat-growing areas, depending on the weather conditions, the grade and condition of the wheat when stored, and the length of time the grain is to be held; but, in general, it may be said to be about 13 to 13.5 percent in the hard red winter wheat area, 13.5 to 14 percent in the soft red winter wheat areas and 14 to 14 ½ percent in the hard red spring wheat area

The external evidences of wheat deterioration usually are not apparent until the condition of the grain has been lowered enough to affect its commercial grade. Percent germination and fat acidity [3] of sound wheat were found to be valuable indices of incipient deterioration that is not apparent from grade examination. From laboratory and field tests, changes in both germination and fat acidity are found to be accelerated under high moisture and temperature conditions, and these changes are measureable before the development of abnormal odors or kernel damage. Under low wheat moisture and temperature conditions there is little loss in germination or increase in fat acidity. The relations between changes in germination and fat acidity on the one hand and wheat temperature and moisture content on the other are discussed at greater length on page 14, in connection with figures 8 and 9.

Aside from damage caused by heat from wheat respiration and micro-organisms, insect infestation may, in addition to the damage caused by the actual feeding upon the kernel, produce enough heat and moisture to start the self-accelerating process mentioned before, and should be kept under control by proper fumigation.

MEANS OF PREVENTING DAMAGE IN STORED WHEAT

As stated before, damage in stored wheat is usually caused by heat from the respiration of micro-organisms or from too rapid respiration of the wheat kernel itself. This damage can be prevented by lowering the moisture content enough to retard these forms of respiration, at the same time dissipating any heat produced in the mass. In designing structures for the safe storage of damp wheat, two problems must be solved—the dissipation of heat, and the removal of moisture from the wheat. Both must be accomplished before material damage has occurred to the wheat.

DISSIPATING EXCESS HEAT

The excess heat evolved in a bin of wheat may be transferred from the wheat mass to the bin wall by conduction or radiation, and then to the surrounding air, or may be removed by convection currents of air.

Tests made in connection with this investigation by the National Bureau of Standards (P. 73) indicate that for wheat with a moisture content between 12.5 and 14 percent the thermal conductivity of the wheat varied from 0.89 to 0.98 British thermal units per hour per square foot per inch thickness. This means that wheat as an insulator is about as effective as sawdust, and will oppose the transfer of heat from the mass to the bin wall. Many temperature readings taken by means of thermocouples in the wheat mass have shown that, where conduction and radiation alone are depended upon, the removal of heat from a large bin of damp wheat is not fast enough to prevent the building up of temperatures dangerous for safe storage. The amount of heat transfer by conduction may be increased by shortening the distance of heat travel, and this method is used in some types of bin-ventilating systems. However, for rapid removal of heat from large masses, air flow through the grain is necessary.

---

[3] The fat-acidity determination is a chemical test based upon the amount of free fatty acids in the wheat Fat acidity is determined by the number of milligrams of potassium hydroxide required to neutralize the free fatty acids in 100 grams of wheat.

EXPOSURE FACTORS INFLUENCING WHEAT MOISTURE

Various investigations have shown that when wheat was exposed to certain constant conditions of relative humidity and temperature, it arrived at rather definite corresponding moisture contents.

Table 1 shows the results of such determinations by Coleman and Fellows within a relatively narrow range of temperatures, which are, however, comparable to those found in most sections at the time wheat is first stored. While no conclusive evidence has been found on the effect of a wide range of temperatures on the moisture content of wheat at any given relative humidity, data on other hygroscopic materials, particularly wood, show that differences in temperature produce noticeable modifications in the relation between relative humidity and moisture content. Limited data obtained in experiments now being conducted as part of this project indicate that the moisture content of wheat reacts to temperature changes in somewhat the same way as does that of wood.

TABLE 1.—*Moisture contents* [1] *of certain classes of wheat in equilibrium with air of various relative humidities at temperatures of 77° to 82.35° F.*[2]

Wheat class	Moisture content of wheat in equilibrium with relative humidities of—						
	15 percent	30 percent	45 percent	60 percent	75 percent	90 percent	100 percent
	Percent	*Percent*	*Percent*	*Percent*	*Percent*	*Percent*	*Percent*
Soft red winter_____	6.31	8.64	10.57	11.88	14.59	19.73	25.63
Hard red winter_____	6.45	8.51	10.52	12.48	14.63	20.10	25.35
Hard red spring_____	6.77	8.47	10.08	11.81	14.76	19.75	25.05

[1] On wet basis.
[2] From the following: COLEMAN, D. A. and FELLOWS, H. C. HYGROSCOPIC MOISTURE OF CEREAL GRAIN AND FLAX SEED EXPOSED TO ATMOSPHERE OF DIFFERENT RELATIVE HUMIDITY. Cereal Chem. 2: 275–287, illus. 1925.

These experiments indicate that at low relative humidities the moisture content of wheat tends to increase as the temperature approaches the freezing point, though at relative humidities of about 75 to 80 percent this tendency appears to be less.

When wheat is completely exposed to air of constant temperature and relative humidity, its temperature and moisture content both presently reach a condition of equilibrium with the surrounding atmosphere, and the moisture content of the wheat, when it reaches such a condition, is the equilibrium moisture content for the given conditions of temperature and relative humidity. Grain of higher or lower moisture content, exposed to the same air conditions, will either give up moisture to, or take up moisture from the air, until it reaches the equilibrium moisture content.

Factors affecting the rate at which moisture is given up or absorbed by wheat are discussed on page 69. One important factor is the amount by which the moisture content of the wheat exceeds or falls short of the condition of equilibrium with the atmosphere. This is illustrated in figure 2 which shows the change in moisture content of a sample of hard red winter wheat, thoroughly exposed to a constant atmospheric condition of 77° F. temperature and a 50-percent relative humidity with air movement at the rate of 100 feet per minute. As the grain dried out at constant temperature, its moisture content approached a condition of equilibrium with the air. The resultant slowing of the rate of drying is plainly shown by the flattening of the curve. With an equal relative humidity, the rate of drying would have been much slower at temperatures corresponding to out-door winter conditions than at 77° F.

If wheat is well exposed to air that is continually changing in temperature and relative humidity, the moisture content of the wheat also will fluctuate. This is shown in figure 3' which is based on the change in weight of a sample of soft red winter wheat (about three kernels deep) in a shallow screen tray which was set on a balance and weighed at 1-hour intervals and on moisture tests with a standard electric moisture meter of samples about six kernels deep on an

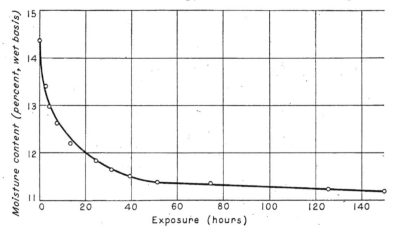

FIGURE 2.—Changes in moisture content of hard red winter wheat as related to time of exposure to air at 77° F. and 50-percent relative humidity.

adjacent screen panel. The moisture tests were made at frequent intervals during the period of exposure. Duplicate samples were immediately stored in sealed cans which were held for 24 hours, before being tested for moisture. The sample used for determining changes in weight was a duplicate of that used for making the initial moisture test. At the conclusion of the experiment, the sample that had been exposed on the balance was used for making the final moisture test. These data show that the moisture in wheat which is almost completely exposed to outdoor air is always approaching an amount corresponding to the equilibrium conditions for each successive relative humidity during the day, giving up moisture when the vapor pressure of the moisture in the wheat is greater than that of the moisture in the air, and absorbing moisture when the conditions are reversed.

Each such gain or loss of moisture is accompanied by heat losses or gains in the wheat kernels, for as moisture evaporates it removes heat. When moisture is absorbed by wheat it gives up corresponding amounts of heat to the kernels. During drying such heat losses from the kernel are to some extent offset by the heat of respiration, by heat flow between the wheat and the ventilating air, and by heat conduction through the wheat mass and the bin floor and walls. However, because of the changes in both moisture content and temperature of the wheat resulting from moisture transfer, the drying process slows down as drying proceeds.

While it is of great importance in designing ventilation systems to know the rate at which moisture is gained or lost by wheat under definitely known or ideal conditions of exposure, it is of equal importance to know what conditions of exposure exist for the wheat that is stored in bulk. Practical storage conditions for wheat do not permit anything approaching complete exposure. That daily fluctuations of weather conditions do not penetrate far into a mass of wheat is shown by continuous records of temperature and relative humidity obtained from carefully screened hygrothermographs buried about 1½ feet below the surface and 1½ to 2 feet from the exposed south wall in bins

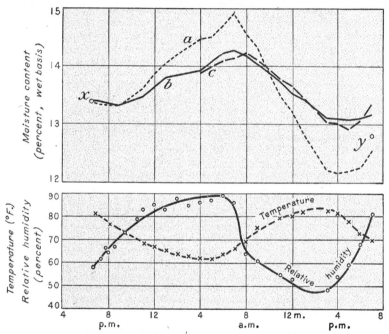

FIGURE 3.—Changes in moisture content of wheat exposed in thin layers: *a*, Computed moisture content based on weight of samples; *b*, moisture tests made during progress of experiment; *c*, moisture tests made after holding duplicate samples 24 hours; *x*, initial moisture content of portion of test sample; *y*, final moisture content of test sample, for which weights are shown in *a*. Circles indicate psychrometer readings.

of various types filled with wheat of known initial moisture contents. The combined records from three of these bins are shown in figure 4 where B–2 refers to an unventilated metal bin; B–6 to a wood bin having horizontal flues spaced 2 feet apart in each direction and open at alternate ends, and B–8, to a wood bin having horizontal flues spaced the same as in B–6 but open at both ends. Complete descriptions of these bins will be found under the 1936 tests at College Park, Md. (p. 9).

It is evident that the relative humidity and temperature of the air contacting the wheat in the depth of the stored mass are determined by the moisture content and temperature of the wheat rather than by the relative humidity and temperatures of the outdoor air. The flues in the ventilated bins used in these tests did not permit the outdoor air to penetrate far enough into the wheat to materially affect the condition of the air surrounding the hygrothermographs. Since the moisture content of wheat and the relative humidity of the space in which it is stored are so closely interrelated, a great variety of means for aerating wheat stored in bulk have been tested in this investigation.

Three general methods of ventilating farm bins are (1) by natural air movement (convection currents) caused by the slight difference in

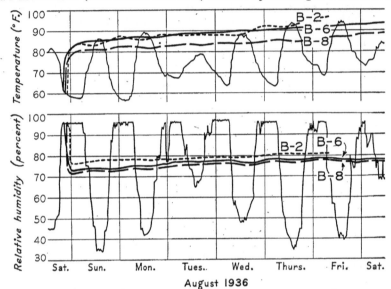

FIGURE 4.—Air temperatures and relative humidities recorded by hygrothermographs buried in wheat, as compared to outside conditions. B-2, an unventilated bin, bins B-6 and B-8 ventilated by horizontal flues spaced 24 inches horizontally and vertically. Each bin contained about 250 bushels of wheat of 16.5 percent moisture content.

density between warm and cool air; (2) by air movements caused by wind pressures; and (3) by forced ventilation using a power-operated blower or fan. Data on the rate of air flow through wheat, the amounts of moisture removed from wheat in bins under laboratory conditions, the relation of wind velocity to static pressures and other information related to design of ventilation systems are presented in the Appendix.

VENTILATION STUDIES IN THE SOFT RED WINTER WHEAT AREA [4]

Soft red winter wheat is grown generally from the Mississippi River eastward to the Atlantic coast. Storage problems for this

---

[4] Valuable assistance in planning and reviewing the work under this project was given by J. E. Metzger, acting director, and by R. W. Carpenter and G. J. Burkhardt of the agricultural engineering staff of the Maryland Agricultural Experiment Station. Mr. Burkhardt also took an active part in the conduct of the work in 1937 and 1938. M. A. R. Kelley, agricultural engineer, and B. M. Stahl, associate agricultural engineer, represented the Bureau of Agricultural Engineering in 1936; B. M. Stahl in 1937 and 1938. Analyses for grade of the samples taken were made by the Baltimore, Md., office of the Grain Division, Bureau of Agricultural Economics.

area may be roughly divided, on the basis of climatic conditions and handling practices, into those of the inland States west of the Appalachian Mountains including Illinois, Michigan, Indiana, Ohio, and Iowa, and those of the mountain and coast States, New York, Pennsylvania, Maryland, and Virginia. Maryland, with wheat-growing sections both in the mountains and on the Coastal Plain, was considered to be representative of the eastern area, and Illinois of the inland States.

MARYLAND TEST, 1936

Because the facilities of the Maryland Agricultural Experiment Station were available at College Park, Md., with temperature, humidity, and wind-velocity conditions about as adverse to successful

FIGURE 5.—Experimental bins at College Park, Maryland. Ventilated bins B–6 and B–8 are the west halves of the two low wooden bins in the background. B–23 and B–24 are the tall, narrow cylinders by the light pole.

storage as at any location in the State, the experimental bins were placed there (fig. 5). The available site was on low ground surrounded by trees or buildings to within a distance of 200 feet, which tended to shut off any wind and also probably increased the average temperatures and humidity.

In 1936, two naturally ventilated bins, B–6 and B–8 were set up at the test location. Both were of wood, 5 by 10 feet in floor area and 8 feet high to the plates, with wood roofs covered with roll roofing. Each was the west half of a two-bin, gable-roofed structure. The ventilating system installed in these bins had been used with some success by the North Dakota Agricultural Experiment Station, and consisted in each case of six horizontal flues each made of 1- by 4-inch boards set on edge 4 inches apart and covered on the upper surface with galvanized fly screen. The six horizontal flues were arranged in two vertical rows of three each, running north and south, or the long way of the bins, and 18 inches from the east and west walls. This made the horizontal distance between flue centers 24 inches. The vertical distance between flue centers was 24 inches, with the bottom ones 18 inches above the floors. The cost of material for these flues when so spaced is about 6 to 8 cents per bushel of bin capacity. The flues in B–8 were open to the air on both ends, while in B–6, alternate ends were closed, the theory being that air entering the open end would be forced by wind pressure through the wheat and escape through a flue with the opposite end open. There were also small screened openings in the gable ends of each bin, allowing a free cir-

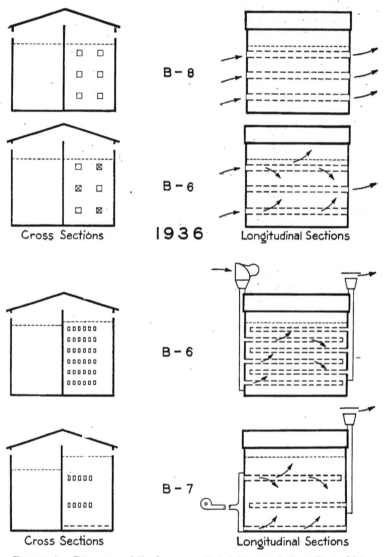

FIGURE 6.—Diagrams of the large ventilated bins at College Park, Md.

culation of air over the top of the grain. The ventilating systems are shown diagrammatically in figure 6. An unventilated bin, B-7, the same size and shape as B-6 and B-8, was used as a check bin.

On July 11, 1936, each bin was partly filled with 255 bushels of soft red winter wheat that had been well mixed, making a depth 6½ feet. The average initial wheat moisture content in B-6 was 16.5 percent, and in B–7 and B–8, 16.4 percent, and the temperature of the wheat when placed in the bins averaged 86° F. The wheat temperature in the center of the unventilated check bin B–7 increased to 120° after 32 days' storage. An average sample taken on August 13 showed the wheat to be sour and heating. The wheat temperatures in the center of each ventilated bin increased steadily until, after 41 days in storage in B–6 it reached 112°, and in B–8, after 39 days in storage, it reached 113°. These temperatures were taken within 8 inches of the ventilating flues. The outside temperatures then began to drop, as shown in figure 7, and the bins cooled. On September 17, after 68 days in storage, an average sample was drawn from each of the ventilated bins and the bins were emptied. The wheat when removed did

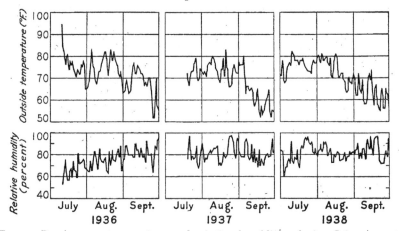

FIGURE 7.—Average temperature and relative humidities during July, August, and September from hygrothermograph records at College Park, Md.

not flow freely, but there was no caking around the flues. The grade-factor data, percent germination, and fat acidity of the wheat before and after storage are given in table 2. The average daily temperature and relative humidity, as recorded by a hygrothermograph at the site, are given in figure 7.

There was no difference in the amount of moisture (1.2 percent) removed from bins B–6 and B–8, and the fact that there appears to be a difference in favor of B–8 in regard to loss in germination or increase in percent damage, might be attributed to the fact that unventilated bin B–7, adjacent to B–8, was emptied twice for cooling and this may have resulted in lower average temperatures for B–8, which could not be directly credited to the ventilating system. The unventilated bin adjacent to B–6 also heated but was not moved for cooling and some of the heat from that bin probably passed through the partition to the wheat in B–6.

Although the two ventilated bins did not heat so quickly as did the unventilated bin filled with wheat of the same moisture content, these types of ventilation cannot be considered effective in storing wheat of moisture content approaching 16.5 percent in the Maryland area. For studies of ventilation with flues of this type in other areas, see the discussion of the Urbana, Ill., 1936 tests and the Fargo, N. Dak., 1936, 1937, and 1938 tests.

TABLE 2.—*Grade-factor data, percent germination, and fat acidity of Maryland 1936 and 1937 wheat samples*

BIN B-7 (UNVENTILATED CHECK BIN)

Sample	Date	Mois- ture	Test weight per bushel	Dam- age	Grade [1]	Odor	Germi- nation	Fat acidity
		Percent	*Pounds*	*Percent*			*Percent*	*Units*
Initial_____	July 13, 1936	16. 4	59. 0	0. 2	Sample [2]_____	Natural_____	83	[3] 17. 4
Final_____	Aug. 13, 1936	16. 4	56. 8	_____	_____do_____	Sour_____	48	_____

BIN B-6 (HORIZONTAL FLUES, ALTERNATE ENDS CLOSED)

| Initial_____ | July 13, 1936 | 16. 5 | 59. 0 | 0. 4 | Sample [2]_____ | Natural_____ | 94 | [3] 17. 6 |
| Final_____ | Sept. 17, 1936 | 15. 3 | 54. 0 | 84. 0 | Sample, Tough _ | Sour_____ | 0 | _____ |

BIN B-8 (HORIZONTAL FLUES. BOTH ENDS OPEN)

| Initial_____ | July 13, 1936 | 16. 4 | 59. 0 | 0. 2 | Sample [2]_____ | Natural_____ | 83 | [3] 17. 4 |
| Final_____ | Sept. 17, 1936 | 15. 2 | 55. 6 | 52. 0 | Sample, Tough__ | Sour_____ | 41 | _____ |

BIN B-26 (UNVENTILATED CHECK BIN)

| Initial_____ | July 21, 1937 | 15. 2 | 57. 8 | 1. 8 | 3 Tough_____ | Natural_____ | 90. 7 | 14 |
| Final [4]_____ | Apr. 23, 1938 | 15. 3 | 57. 5 | 2. 8 | Sample, Tough__ | Sour_____ | 66 | _____ |

BIN B-6 (WIND-ASSISTED NATURAL VENTILATION)

Initial_____	July 21, 1937	15. 1	57. 8	1. 8	3 Tough_____	Natural_____	92. 5	15
Average_____	Apr. 23, 1938	14. 3	58. 2	2. 0	2 Tough_____	_____do_____	78. 2	_____
Latest_____	Mar. 27, 1939	13. 6	58. 0	2. 9	2_____	_____do_____	58. 0	26. 2

BIN B-7 (FORCE-VENTILATED)

Initial_____	July 19, 1937	15. 1	57. 6	1. 7	3. Tough_____	Natural_____	86	19
Average_____	Apr. 23, 1938	14. 0	58. 2	2. 0	2_____	_____do_____	71. 5	_____
Latest_____	Mar. 27, 1939	13. 5	58. 0	3. 0	2_____	_____do_____	39. 5	29. 4

[1] All wheat garlicky and of class soft red winter, subclass red winter. Dockage not over 0.3 percent in any sample.
[2] Sample grade because of excess moisture.
[3] Fat acidity calculated from acid value determination and average oil content of soft red winter wheat.
[4] First sample taken after wheat went out of condition.

MARYLAND TESTS, 1937 AND 1938

As pointed out in the Appendix (p. 66) little power is available from the wind, and the amount of air forced through wheat at a given pressure, as well as the drying rate, is increased by shortening the length of the air path through the wheat—in other words, placing the fiues closer together. The bins were remodeled in 1937 along these lines. The six fiues of B–6 were replaced with six layers or grids of, fiues with a vertical distance between layers of 10 inches, as shown in figure 6. Each layer was formed by a series of 1-by 2-inch by 10-foot pieces set on edge 3 inches on centers, with alternate spaces between the pieces covered top and bottom with galvanized fly screen. The cost of material for such a system will range from 15 to 25 cents per bushel, being lower for the larger bins. The layers were 10 feet long but only 3 feet wide, leaving an unventilated area 2 feet wide at the west wall. The first, third, and fifth layers from the bottom were connected at one end to a pressure-type ventilating cowl, kept headed into the wind at all times by a vane. The other three layers were connected to a suction cowl. The air path was from the pressure cowl to the horizontal flues, through about 1 foot of wheat, and out the suction cowl via the three layers of flues to which it was connected. The six layers of horizontalyflues had a total screened area of 130.6 square feet, divided equally between inlet and outlet flues.

Bin B–7, the east half of another structure similar to that housing B–6, was used for forced-ventilation studies. A floor of perforated metal was laid on 2 by 4's over the wooden floor, leaving an air chamber 3¾ inches deep and approximately the size of the floor area (5 by 10 feet). Two layers of horizontal flues, constructed the same as those in B–6, were installed, one 2 feet and the other 4½ feet above the floor. The lower grid was connected to a suction-type ventilating cowl above the bin and the upper grid and the floor air chamber to a small multivane blower operated by a ¼-horsepower electric motor. A diagram of this system is shown in figure 6.

As a check on the performance of these two ventilated bins (B-6 and B-7), bin B-26, the west half of a 500-bushel round, metal bin with metal floor was used. The bin had no roof of the conventional type, but was sheltered from sun and rain by an open shed with corrugated metal roof and walls on the east and west ends. This provided shade during the hottest hours of the day and allowed free circulation of air over the surface of the grain.

Bins B–6 and B–7 were filled on July 19, 1937, with 266 bushels each of 15.1-percent moisture content soft red winter wheat purchased on the terminal market at Baltimore. B–26 was filled with 180 bushels of wheat from the same lot but which averaged 15.2 percent moisture on July 20. Samples were taken from each bin once each week for several weeks, and then at longer intervals for moisture, grade, germination, and rancidity determinations. Temperatures at five different points in each bin were observed three times each day during the first 2 months, and thereafter at longer intervals.

The wheat was still in storage in the naturally ventilated bin, B–6, on April 1, 1939, the pressure and suction cowls being in place during the entire period. The blower was operated in B–7 only during the first 21 days of storage, at times when the air relative humidity was below 60 percent. Sixty percent relative humidity was taken as the maximum for drying because, as stated in the appendix, little drying was accomplished in the laboratory experimental bin at normal temperature when this figure was exceeded. The blower, when in operation, forced approximately 393 cubic feet of air per minute through the wheat. The complete schedule of blower operation is shown in table 3.

The wheat in the two ventilated bins, B–6 and B–7, was still in good condition after 614 days storage, and increased one numerical grade due to an increase in test weight. However, the germination had fallen and the fat acidity increased. The wheat in the unventilated bin, B–26, turned sour after 276 days in storage.

The grade factor data, percent germination, and fat acidity of the wheat in these three bins, before and after storage, are shown in table 2.

The mean daily temperature and relative humidity for the storage period, as recorded at the bin site, are given in figure 7.

TABLE 3.—*Schedule of blower operation, bin B–7*

Dates operated, 1937	Time operated	Total air volume[1]	Average air condition during operation		Dates operated, 1937	Time operated,	Total air volume[1]	Average air condition during operation	
			Temperature	Relative humidity				Temperature	Relative humidity
	Hours	*Cubic feet*	*°F.*	*Percent*		*Hours*	*Cubic feet*	*°F.*	*Percent*
July 19	1¾	41,000	83	56	Aug. 1	7½	177,000	82	56
21	2½	59,000	89	58	2	9	212,000	83	49
22	8	188,000	82	54	3	5	118,000	85.5	50
23	6	141,000	86	54	4	7½	177,000	85	48
24	11⅙	263,000	82	54	5	6½	153,000	87	54
25	7¼	172,000	87	54	6	5½	130,000	87.5	55
26	1⅚	42,000	85	62	7	4¾	112,000	88	56
27	9½	224,000	80	43	8	6½	153,000	88.5	46
28	9¾	232,000	80	39	9	4½	106,000	88	57
29	8	188,000	83	47					
30	5	118,000	85	55	Total	----------	3,006,000	----------	----------
31	0	----------	----------	----------					

[1] Air volumes were calculated from anemometer readings taken on only 1 day, and are approximate only.

Wheat moisture content, temperature, percent germination, and fat acidity are shown for the three test bins in figures 8 and 9. Comparison of the amounts of moisture removed from the two ventilated bins B–6 (natural ventilation) and B–7 (forced ventilation) during the storage period, from July 21 to August 9, the period in which the blower was operated intermittently, shows that the amount removed by forced ventilation was about twice that removed by natural ventilation. The 3,006,000 cubic feet of air forced through B–7 removed on the average 0.23 grain of moisture per cubic foot. Samples taken during the period of forced ventilation from different parts of the bin indicated that practically no moisture was removed from the wheat near the outlet flues. After operation of the blower was discontinued, drying progressed in the two bins at the same rate up to September 30, 1937, as shown in the curves of figure 8. From September 30, 1937, until the last of April 1938, no moisture was lost from either of the ventilated bins, but after this date there was a steady drop in bin B–6 until October 1, 1938, when drying again stopped, probably because of the increase in humidity of the ventilating air. During the same period, B–7 lost moisture at about half the rate of B–6. After operation of the blower was discontinued in bin B–7, it became a naturally ventilated bin, but because of the greater distance between flues and because it had no pressure cowl to assist the air through the wheat, it was not as well ventilated as B–6.

The average temperature of the ventilated bins followed the outside temperatures more closely than did that of the unventilated bins. Cooling was quickest in naturally ventilated bin B–6, probably because of the thin layers of wheat.

Changes in germination and fat acidity were affected by both wheat moisture content and wheat temperature. After September 30, when the wheat temperatures went below 65° F., little change in germination and fat acidity was noticed in any of the bins. However, as soon as the temperature started to rise in the spring (about April 1), the rate of change in both germination and fat acidity increased rapidly in all bins, being greatest in B–26, the unventilated bin. It was at this time that the wheat in bin B–26 turned sour, and during the summer months it lost all germination and increased in fat acidity to 45.7 units on September 16, 1938.

The temperature of the wheat may change the initial condition of the ventilating air with respect to temperature and relative humidity; and cool, damp air passing through wheat of a higher temperature may, under certain conditions, remove just as much or more moisture than relatively dry air forced through cold wheat. These conditions could occur in the case of night ventilation. To compare the effectiveness of night ventilation in relation to day ventilation in naturally

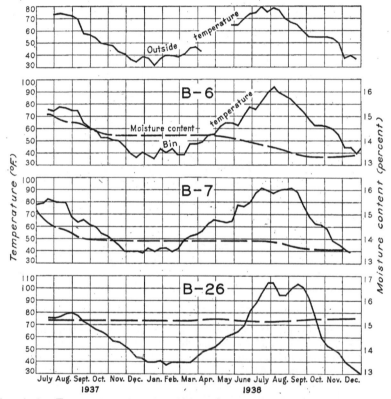

FIGURE 8.—Temperature at center of bin and average moisture content of wheat in ventilated bins B–6 and B–7, and nonventilated bin B–26, and corresponding outside air temperature from July 1937 to December 1938.

ventilated bins, two small, circular, metal bins, 27 inches in diameter and 8 feet high, were equipped with seven layers of horizontal flues, spaced approximately 1 foot apart, vertically. The second, fourth, and sixth layers from the bottom in each bin were connected to a suction cowl. The other flues were connected to inlet ducts open to the air near the bin bottoms, and provided with tight-fitting covers so that the air circulation through the wheat could be entirely shut off when desired. On July 29, 1937, both bins were filled with 15.5-percent-moisture wheat from the same lot. In one bin night ventilation was provided by opening the inlet duct between 8 p. m. and 7 a. m., and on the other, day ventilation, by leaving the inlet duct

open between 7 a. m. and 8 p. m.   There was no significant difference
in the rate of drying in the two bins.   As the wind velocity is usually
much lower at night (averaging 0.9 mile per hour as compared with
3.3 miles per hour during the day, in the summer months of 1936
when day and night records were kept) there was probably very
little air drawn through the wheat during the night hours.

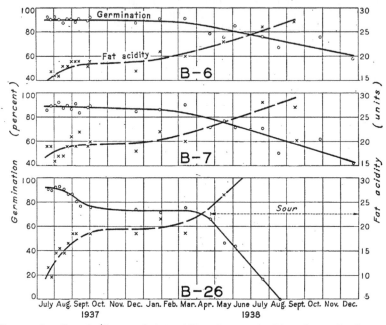

FIGURE 9.—Germination and fat-acidity changes in bins shown in figure 8.

RESULTS OF MARYLAND TESTS

In comparing the results of the 3 years' tests of ventilated bins at
College Park, the weather prevailing during the three summers and
the initial condition of the various lots of wheat must be considered.

The most dangerous season for keeping wheat in Maryland is the
summer, for both the temperature and the relative humidity are
normally higher than at any other season.

The most critical month is August, for wheat is generally placed in
storage during the first 2 weeks of July when summer temperatures
prevail, and unless the moisture content of the wheat is reduced
during July, one or more short periods of maximum summer temperatures, such as often occur during August will cause the development of musty or sour odors or at least a progressive increase in fat
acidity, depending on the amount of moisture in the wheat when
these high-temperature periods appear.   If the moisture content of
the wheat is low enough to store safely during the first summer it
may go out of condition during the second summer following the
warm spring weather that tends to build up a higher wheat temperature with a corresponding stimulation of any micro-organisms or
insects that are present.

Data from Weather Bureau records at Washington, D. C., the nearest station to College Park, are given in table 4. The departures from normal indicate that the summer of 1936 was somewhat warmer and drier than average for this locality, though the mean relative humidities were higher than in the three other locations where ventilation studies were made. Comparisons between the ventilated and the nonventilated bins at College Park were more than ordinarily favorable to the ventilated bins in 1936. In 1937, July was almost an average month while August presented less favorable conditions than normal, and September was slightly more favorable. The summer of 1938 was slightly less favorable for wheat storage than the average summer.

TABLE 4.—*Mean summer temperatures and relative humidities and departures from normal at Washington, D. C., 1936–38*

Year	July				August				September			
	Mean		Departure from normal		Mean		Departure from normal		Mean		Departure from normal	
	Temperature	Relative humidity	Temperature	Relative humidity	Temperature	Relative humidity	Temperature	Relative humidity	Temperature	Relative humidity	Temperature	Relative humidity
	°F.	Percent	°F.	Percent	°F.	Percent	°F.	Percent	°F.	Percent	°F.	Percent
1936	78.4	68	+1.6	−6	77.9	71	+2.9	−4	71.4	74	+3.3	−3
1937	77.2	72	+.4	−2	77.8	79	+2.8	+4	65.9	77	−2.2	0
1938	78.4	75	+1.6	+1	78.6	73	+3.6	−2	67.4	78	−.7	+1

As indicated in table 2, the initial moisture contents of wheats stored in the ventilated bins in 1936 were 16.4 and 16.5 percent. The storage hazard due to the higher moisture content of this wheat as compared with the wheats of 15.1 and 15.2 percent moisture stored in 1937 much more than offset the effect of the favorable season in 1936. However, in wet seasons there is much wheat threshed in Maryland with moisture contents of 16 percent or more. The result of studies indicate that ventilation as a means of reducing the moisture content of wheat stored in bulk is comparatively difficult in a climate such as at College Park, during July, August, and September. However, conditions for drying during the winter are relatively favorable and if wheat can be kept from heating until cool weather it can be dried out slowly in ventilated bins before the next warm season.

The horizontal flues with 24- by 24-inch spacing were not effective in preventing heating of wheat of 16.4- and 16.5-percent moisture content in 1936, but probably would have kept wheat of 15-percent moisture. The closely spaced flues attached to pressure and suction cowls, as used in bin B–6, were very effective in preventing heating and removing moisture, and would probably have succeeded in holding sound wheat of 16-percent moisture. Forced ventilation during the driest parts of the days with air inlets and outlets arranged to provide escape for the air after travel through about 2 feet of wheat, was also quite satisfactory, but might not be dependable in seasons of higher relative humidity.

VENTILATION STUDIES IN ILLINOIS, 1936 [5]

Because of the facilities of the University of Illinois available at Urbana, that location was selected for investigations of wheat storage in the soft red winter wheat area west of the Appalachian Mountains. A series of corncribs, of about 300 bushels capacity each, and mounted on flanged wheels so that they could be moved over a track to scales for weighing, were remodeled for the wheat-storage tests. The track, scales, and movable cribs were in a well-drained location, exposed to the wind from the north, west, and east but sheltered by large buildings about 100 feet to the south. They were exposed to the sun from all directions, but shaded each other in the early morning and late afternoon.

Four cribs were remodeled in 1936 for ventilation tests—two for natural ventilation by horizontal flues, and two for forced ventilation with hot and cold air.

TESTS WITH HORIZONTAL VENTILATING FLUES

The original walls of the cribs were of 1- by 6-inch beveled boards set about 1 inch apart, on the north and south sides, and of 1- by 6-inch tongue-and-groove boards set tight on the east and west ends. These

walls were left in place for the wheat-storage tests and the inside of the studs lined on all sides with tongue-and-groove boards, making a grain-tight bin measuring 7 feet wide (north to south) and 9 feet long (east to west) (fig. 10). The west halves of bins U–3 and U–4 were fitted with six horizontal flues each, of the same type as used in the Maryland bins in 1936. The flues were arranged in three layers of two each, the bottom layer just off the floor, and the second and third layers 2 and 4 feet, respectively, above the floor. The

FIGURE 10.—Movable corncribs remodeled and used for wheat-storage studies at Urbana, Ill.

horizontal distance between flue centers was 13 inches, the west row being 10 inches from the wall. Both ends of all flues in bin U–3 were open to the air. In bin U–4 one end of each flue was closed, alternate ends in each layer being open on the north and south. The flues arranged thus gave natural ventilation to the west halves of both bins but no ventilation to the east halves. Diagrams of these are shown in figure 11.

[5] Valuable assistance in planning and reviewing the work under this project was given by E. W. Lehmann and W. A. Foster of the Department of Agricultural Engineering and W. L. Burlison of the Department of Agronomy, Illinois Agricultural Experiment Station, and by M. D. Farrar, of the State Natural History Survey. The Bureau of Agricultural Engineering was represented by Thayer Cleaver, assistant agricultural engineer. Analyses for grade of the samples taken were made in the St. Louis, Mo., office of the Grain Division, Bureau of Agricultural Economics.

On July 2, bin U–3, with both ends of all flues open, was filled to the 5-foot 8-inch level with 281 bushels of 15.7 percent moisture content soft red winter wheat. The damage at the time of storage, amounting to 2.4 percent, consisted of immature, moldy, and sick kernels.

Both halves of the bin were equipped with thermocouples for temperature determinations. The wheat in the approximate center of

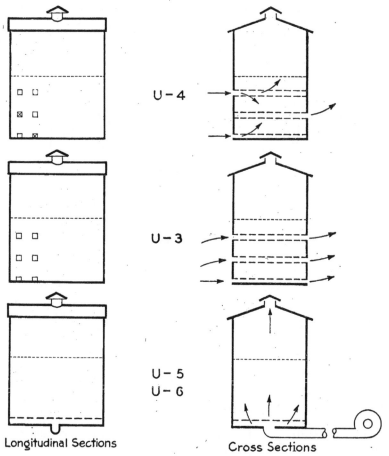

Longitudinal Sections   Cross Sections

FIGURE 11.—Diagrams of ventilated bins at Urbana, Ill., in 1936. Supporting wheels not shown.

mass in each half had an initial temperature of 86° F., rising rapidly in the unventilated half to a high temperature of 120° on July 20. At this time it was moved for cooling, after only 18 days' storage. A sliding partition was inserted from the top between the ventilated and unventilated halves to hold the ventilated wheat in position during the unloading of the heating wheat. During the same 18-day period the highest temperature noted in the ventilated end of the bin was 92°. After the heating wheat was moved the temperature of the ventilated wheat returned to 86°. The wheat in the unventilated end of the bin was moved three times during the fall, each time because of heating, and this added heat undoubtedly affected the damage, germination, and fat acidity of the ventilated wheat which was emptied on September 29, after being in storage 89 days without developing a musty odor. However, the grade, germination, and fat acidity data given in table 5 indicate that the wheat was not in good condition. During the storage period, 1.6 percent moisture was removed from the ventilated wheat.

Bin U–4, remodeled in exactly the same manner as U–3 except that one end of each flue was closed, was filled on July 3, 1936, with 281 bushels of 14.2-percent moisture content wheat. The damage, amounting to 1.2 percent, consisted of green and immature kernels.

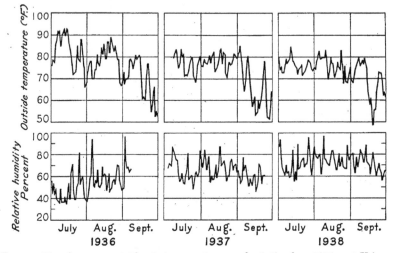

FIGURE 12.—Average outside air temperatures and relative humidities at Urbana, Ill., July, August, and September 1936, 1937, and 1938.

The grain was left in storage until September 23, 1937, a period of 447 days, at which time the wheat in the unventilated end developed a musty odor. The initial and final grade, germination, and fat acidity data are given in table 5.

Thermocouples at the approximate center of mass in each end of the bin showed wheat temperatures of about 90° F. at the time of filling. The unventilated wheat maintained this temperature for 3 weeks and then dropped gradually to 81° on August 15, and, except for slight variations, remained at that temperature until September 29, when the readings were discontinued. The temperature at the corresponding point in the ventilated half of the bin (which was usually the warmest point in that end) dropped from the initial 90° to 82° in 5 days and then ran parallel to, but about 10° lower than, the unventilated wheat, until 72° was reached. On August 10, 1936, the wheat temperature began to rise, due to an increase in the air temperature (fig. 12), reaching a maximum of 85° on August 28, after which it cooled gradually to 68° on September 29. Temperature readings were resumed July 8, 1937, wheat in both ends of the bin being at 72°. The temperature in the unventilated end increased steadily to 95° on September 19, at which time the wheat turned sour. The temperature in the ventilated end increased less rapidly to 80° on August 10, this temperature being maintained until September 7, after which it dropped rapidly to 70° on September 23, 1937.

Only one sampling of the bin was made for moisture determinations during the fall of 1936. On September 29, 1936, the average moisture content of the ventilated end was 12.9 percent and of the unventilated end 14.1 percent. The 1937 moisture contents are given in table 5.

TESTS WITH FORCED VENTILATION, HEATED AND UNHEATED AIR

Two of the movable corncribs, U–5 and U–6, were remodeled for forced ventilation of wheat. The walls were lined with 1- by 6-inch tongue-and-groove flooring, and an air chamber was formed on the bottom of each by laying a floor of perforated metal over 2- by 8-inch joists set on the old, tight, wood floors (fig. 11). These chambers were arranged so that connections could be made, through ducts, to a supply of normal temperature air in the case of U–5, and to a source of heated air in the case of U–6. The same blower was used to supply air to both bins but they could not be ventilated simultaneously.

Bins U–5 and U–6 were filled on July 3, 1936, from the same car of wheat as bin U–4. The grain was hauled by truck from the car, and divided about equally by means of a divider, so that each of the three bins received approximately the same kind and moisture content of wheat. By volume, 300.8 bushels were stored in U–5 and 301.6 bushels in U–6. This filled each bin to a level of 5 feet 11 inches.

Normal outside air was forced into the air chamber of bin U-5 and up through the 5 feet 11 inches of wheat, for 3 hours on each of 7 days, July 6, 7, 8, 9, 10, 11, and 13, making a total ventilation period of 21 hours. The average air temperature was 107° F. The average rate of air flow was 1,885 cubic feet per minute, or a total of 2,375,100 cubic feet. Two hundred forty-three pounds of weight were removed, part of this loss no doubt being moisture from the lumber of the bin. Moisture tests of average samples of the wheat indicated a total loss in moisture content of 0. 6 percent, or only about half that shown by the loss in weight. There was a drop of 0. 5 pound in the test weight of the wheat between July 4 and September 26 as shown in table 5. Samples taken from various locations for moisture determinations showed considerable decreases in the moisture content of the wheat in the lower part of the bin, but there was an increase in moisture content of the wheat in the upper part of the bin during the first few days of ventilation.

The wheat temperature at the time of filling bin U–5 ranged between 82° and 90° F. This remained fairly constant until ventilation was started on July 6, when the temperature in the center of the bin, 7 inches above the floor, increased slowly to 108° on July 11 following the increase in the outside air temperature. When ventilation during the heat of the day was stopped after July 13, all temperatures dropped again to the initial values.

Bin U–6 was ventilated for 3 hours on each of the same days on which U–5 was ventilated—July 6, 7, 8, 9, 10, 11, and 13. The air was heated to an average temperature of 145° F. but at times was much hotter. It was supplied at the same rate as to bin U–5, 1,885 cubic feet per minute. The total loss in weight was 531 pounds and the average moisture content was lowered 1. 2 percent, again only about half that indicated by the weight loss. In this case there was a loss of 1.0 pound in the test weight of the wheat. As in bin U-5, the moisture content of the wheat near the top of the bin increased during the first part of the ventilating period.

The temperature of the wheat 7 inches above the floor of bin U-6, before ventilation with hot air was started, averaged 86° F. The temperature at this level during the ventilation period ranged from 118° to 198°, but with increases of only a few degrees in the wheat temperature at the 3-foot and 5-foot levels. After ventilation with hot air was stopped on July 13, it became necessary to blow normal-temperature air through the wheat in order to lower the temperatures near the floor.

TABLE 5.—*Grade-factor data, percent germination, and fat acidity of Illinois, 1936 wheat samples*

BIN U-3 (WEST HALF VENTILATED WITH HORIZONTAL FLUES, BOTH ENDS OPEN)

Sample	Date	Mois-ture	Test weight per bushel	Dock-age	Dam-age	Grade [1]	Odor	Ger-min-ation	Fat acid-ity
		Pct.	*Lbs.*	*Pct.*	*Pct.*			*Pct.*	*Units*
Initial........	July 3, 1936	15.7	61.0	0.6	2.4	Sample, Light Garlicky.	Natural...	84	[2]15.96
Final (venti-ated end).	Sept. 29, 1936	14.1	59.8	.5	18.0	Sample, Tough, Weevily, Light Garlicky.	...do.......	17	[2]48.52

BIN U-4 (UNVENTILATED EAST HALF)

Initial........	July 4, 1936	14.2	61.6	0.5	1.2	1 Light Garlicky, Tough.	Natural...	84	[2]17.30
Average......	Sept. 29, 1936	14.10	60.6	.5	1.6	1 Light Garlicky, W e e v i l y, Tough.	...do.......		[2]29.48
Do.........	Aug. 4, 1937	14.6	59.8	.4	7.0	3 Tough............	Slight weevily	44.7	[2]37.0
Final.........	Sept. 23, 1937	14.2	59.0	.5	12.2	Sample, Tough, Weevily.	Musty....	18.7	34.0

BIN U-4 (WEST HALF VENTILATED WITH HORIZONTAL FLUES, ONE END ONLY OPEN)

Initial........	July 4, 1936	14.2	61.6	0.5	1.2	1 Light Garlicky, Tough.	Natural...	84	17.30
Average......	Sept. 29, 1936	12.9	60.6	.4	1.7	1 Light Garlicky, Weevily.	...do.......		[2]22.2
Do	Aug. 4, 1937	13.7	59.7	.5	4.5	3...............	Slight weevily	69.7	25.0
Final.........	Sept. 23, 1937	13.59	59.5	.5	4.6	3 Weevily.........	Weevily...	59.7	27.0

BIN U-5 (FORCE VENTILATED FROM BOTTOM WITH NORMAL AIR)

Initial........	July 4, 1936	14.2	61.4	0.3	1.0	1 Tough............	Natural...	86	[2]17.8
Average......	Sept. 29, 1936	13.5	60.9	.4	1.5	1 Weevily............			[2]24.96
Final.........	June 15, 1937	14.7	60.0	.3	6.0	3 Tough............	Natural...	26	......

BIN U-6 (FORCE VENTILATED FROM BOTTOM WITH HEATED AIR)

Initial........	July 4, 1936	14.2	61.4	0.3	1.5	1 Tough............	Natural...	86	[2]15.30
Average......	Sept. 29, 1936	13.9	60.4	.5	1.7	1 Weevily............			[2]21.92
Final.........	June 15, 1937	14.7	59.8	.5	12.8	5 Tough............	Natural...	27	......

[1] All wheat of class soft red winter and subclass red winter.
[2] Fat acidity calculated from acid value determination and average oil content of soft red winter wheat.

The wheat in bins U-5 and U-6 was removed on June 15, 1937, at which time the wheat moisture content had increased to 14.7 percent in both bins. Due to increase in percent damage, the wheat in bin U-5 dropped to grade 3 Red Winter Tough and in bin U-6 to grade 5 Red Winter Tough. The grade, germination, and fat-acidity data of average samples drawn from these two bins are shown in table 5, along with similar data for bins U-3 and U-4.

ILLINOIS TESTS, 1937

TESTS WITH HORIZONTAL VENTILATING FLUES

The 1936 investigations indicated that although the rate of drying obtained with horizontal flues was very slow, benefit was derived from the dissipation of heat. Because of the possibility that the heating wheat in the unventilated ends of bins U-3 and U-4 in 1936 may have affected the wheat temperature in the ventilated ends, the bins were remodeled in 1937 to provide ventilation of the entire mass of wheat. Ten horizontal flues, running north and south, were installed in bin U-3 in 2 layers of 5 flues each, placed 2 and 4 feet, respectively, above the floor. Each flue in the upper layer was directly over the one below it, and the horizontal distance from the end walls to the flues, and between flues, was 1½ feet. One end of each flue was closed, alternate flues being open on the north and south. Bin U-8, another of the 300-bushel portable bins, was equipped with 30 flues, arranged in 6 layers of 5 flues each, with horizontal distances the same as in U-3. The vertical distance between the floor and the bottom layer and between each two layers was 1 foot. As in U-3, one end of each flue was closed, alternate north and south ends being open (fig. 13).

Bins U-3 and U-8 were filled on July 11 with soft red winter wheat, 284 bushels being placed in U-3 and 305 bushels in U-8. The wheat was well mixed but had a moisture content of about 17 percent which was higher than desired for the test. The average temperature of the wheat in both bins when filled was about 89° F. This rose in the center of bin U-3 to a maximum of 100° on July 26. After 25 days of storage, the wheat turned musty and the bin was emptied on August 5. The temperature in U-8 dropped to approximately 80° in 10 days and held at that temperature until, after 40 days in storage, the wheat turned musty and the bin was emptied. (See table 7 for the initial and final grade, fat-acidity, and germination data of these two bins.) When emptied, some wheat in both bins stuck to the walls because of mold, but there was no caking around the flues.

TESTS WITH FORCED VENTILATION

Experiments with forced ventilation in 1936 indicated that much of the ventilating air merely transferred moisture in the mass and did not remove it. It appears that the air forced into and through the moist wheat took up moisture but as this moisture-laden air moved farther through the bin, temperature was lowered by its contact with cooler wheat, and part of the moisture was deposited by condensation on the colder kernels.

To determine the effect on moisture transfer within bins of shortening the length of air travel through the wheat when using forced ventilation, two bins, U-5 and U-6, were each equipped with perforated metal floors on 2- by 6-inch joists laid on the regular wooden floors, so as to provide an air chamber under the wheat, and each was further equipped with two layers of five horizontal flues each, the sizes and horizontal spacing of these flues being the same as in bins U–3 and U–8, already described. In bin U–5 (fig. 13), the lower layer of flues was 2 feet and the upper layer 4 feet above the perforated floor, while in bin U–6 these distances were 1½ feet and 3 feet, respectively.

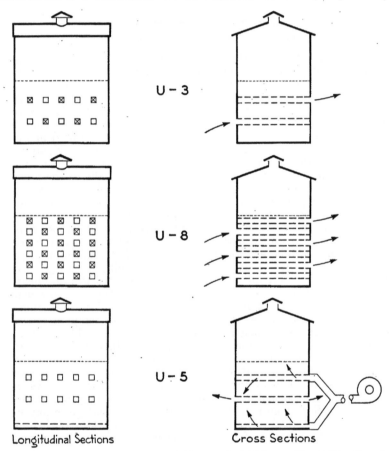

Longitudinal Sections   Cross Sections

FIGURE 13.—Diagrams of ventilated bins used at Urbana, Ill., 1937. Running gears not shown.

Through suitably connected header boxes, air could be forced into the wheat by way of the perforated floor and the upper layers of flues and escape through the lower flues and the upper wheat surface. In bin U–5 the length of the air path through the wheat, from inlet to outlet, was 24 inches while in bin U–6 the length of the air path was 18 inches in the lower part of the bin and 30 inches in the upper part because the flues were placed lower in the bin.

Bins U–5 and U–6 were filled on July 16, 1937, with damp wheat which had been used previously for a few days in experiments with insecticides. Both were filled 5 feet 6 inches above the perforated floor. While the two bins were nominally of equal capacity, U–5 required 284 bushel and U–6 276 bushels to fill them to this level.

Simultaneous ventilation of the two bins with normal-temperature air was started on July 20 and continued intermittently through September 15. The periods of blower operation are given in table 6, which also gives the actual moisture content of samples taken at three different elevations in each bin immediately after each period of ventilation, and the initial condition of the ventilating air with reference to its temperature and relative humidity. Both bins had the same depth of wheat, but the upper layer of flues in U–5 was 18 inches below the wheat surface and in U–6 over 24 inches, so there was not an equal distance of air travel within each bin, and more air went through some areas than through others. This was apparent soon after ventilation started, from the differences noted in the drying rates in the various levels as given in table 6.

TABLE 6.—*Moisture reduction by forced ventilation, bins U–5 and U–6, 1937*

Date	Duration of blower operation	Bin U–5 [1] Air volume per minute	Moisture T	M	B	Bin U–6 [2] Air volume per minute	Moisture T	M	B	Temperature Average	Maximum	Minimum	Relative humidity Average	Maximum	Minimum
	Hours	*Cubic feet*	*Per cent*	*Per cent*	*Per cent*	*Cubic feet*	*Per cent*	*Per cent*	*Per cent*	*° F.*	*° F.*	*° F.*	*Per cent*	*Per cent*	*Per cent*
July 17	0		16.4	16.4	16.3	-----	16.0	16.0	16.1	----	----	----	----	----	----
20	5	.766	16.2	16.2	16.2	695	16.2	16.2	16.2	80	81	78	42.7	52.0	38.0
21	5⁵⁄₁₂	766	16.4	16.4	16.4	695	16.2	16.0	16.6	82	83	80	36	40	34
22	5	766	16.45	16.0	16.4	695	16.1	16.0	16.3	84	85	82	40	43	39
23	5	766	15.57	16.3	16.4	695	16.0	15.8	16.1	85	87	82	44	52	40
24	5	766	15.49	15.7	16.7	695	16.4	15.6	16.3	87	88	84	56	61	54
27	5	766	14.77	14.9	16.4	695	16.1	14.9	15.8	78	79	75	38	40	37
28	5	766	14.5	15.1	16.6	695	16.3	15.1	16.1	81	83	78	46	47	43
31	5	766	14.1	15.7	16.3	695	16.0	14.8	15.9	84	85	82	49	52	47
Aug. 2	5¼	605	14.4	15.0	16.5	678	16.0	14.1	16.1	85	87	81	50	56	46
3	5	605	14.2	14.4	16.7	678	16.2	16.0	14.2	87	88	84	.50	54	47
6	5	605	14.9	15.0	16.5	678	16.1	14.5	15.8	82	85	78	54	60	50
7	5	605	13.7	14.1	16.5	678	15.8	13.3	15.2	88	92	82	60	70	54
13	3	506	13.35	13.93	16.13	776	15.51	12.85	15.09	81	83	79	40	42	38
14	5	506	13.87	13.94	15.43	776	15.57	12.48	14.78	83	85	79	43	47	42
17	3	506	13.04	13.9	16.16	776	15.19	12.49	14.7	94	95	92	44	51	41
19	3¾	506	14.98	13.97	13.49	776	14.98	13.25	14.94	88	89	88	41	46	38
25	3¹⁄₁₆	506	13.56	13.79	15.54	776	14.74	11.92	14.0	89	93	85	44	48	41
27	5	506	13.46	14.19	15.27	776	14.46	11.95	14.78	93	97	89	37.5	45	30
Sept. 7	3¾	495	13.48	13.88	14.75	870	14.22	11.6	14.2	81	82	80	34.5	35	34
8	5½	495	13.58	13.95	14.73	870	13.88	11.52	14.07	87.5	91	84	35.5	42	29
9	5	495	13.45	14.03	14.78	870	13.75	12.01	13.75	93	96	90	33.5	36	31
14	4	495	13.34	13.9	14.38	870	13.4	11.25	13.72	72.5	75	70	39	43	35
15	5½	495	13.32	13.95	13.45	870	13.1	11.13	13.37	69	71	67	35	37	33

[1] T samples drawn 4 inches below upper wheat level and 14 inches above upper layer of flues; M samples 9 inches above lower layer, 12 inches below upper layer of flues; B samples from 6 inches above the perforated floor.

[2] T samples 12 inches above upper layer of flues and 12 inches below upper wheat surfaces; M samples equidistant from upper and lower layers of flues; B samples from 6 inches above perforated floor.

Bin U–5 received an average of 620 cubic feet of air per minute or 2.19 per minute per bushel. In the total ventilating period of 107 hours, 10 minutes, 3,988,239 cubic feet of air were forced through the wheat in this bin. This reduced the average moisture content (using the averages of the figures in table 6) from 16.4 percent to 13.6 percent. Calculating from the number of bushels stored and the test weight, a total of 502 pounds of water, or 0.88 grain of moisture per cubic foot of air, was removed. An average of 747 cubic feet per minute, or 2.7 per minute per bushel, was forced into and through bin U-6. By the same method of calculation, the wheat in U-6 was reduced from 16.10 percent to 12.5 percent by 4,807,665 cubic feet of air, which removed a total of 640 pounds of water, or 0.97 grain per cubic foot of air. Unfortunately the track scale on which the bins were weighed was out of order and it was not possible to check calculated against actual weights.

The bins were emptied on September 23, 7 days after ventilation was discontinued because the moisture content of the wheat having been lowered to a safe moisture content for storage, the purpose of the investigation had been accomplished. The initial and final grade, fat acidity, and germination data are given in table 7.

TABLE 7.—*Grade-factor data, percent germination, and fat acidity of Illinois 1937 wheat samples*

BIN U-3 (VENTILATED WITH 10 HORIZONTAL FLUES)

Sample	Date	Moisture	Test weight per bushel	Dockage	Damage	Grade [1]	Odor	Germination	Fat acidity
		Percent	Pounds	Percent	Percent			Percent	Units
Initial	July 12	17.2	56.0	0.5	0.6	Sample, Garlicky	Natural	66.5	19
Final	Aug. 5	16.6	55.0	.5	.9	____do____	Musty	36.5	42

BIN U-8 (VENTILATED WITH 30 HORIZONTAL FLUES)

| Initial | July 12 | 17.0 | 56.0 | 0.5 | 0.7 | Sample, Garlicky | Natural | 69.5 | 21 |
| Final | Aug. 19 | 16.3 | 55.7 | .4 | 2.6 | ____do____ | Musty | 36.7 | 38 |

BIN U-5 (FORCE VENTILATED, AIR PATH THROUGH WHEAT 18 INCHES)

| Initial | July 16 | 16.3 | 54.5 | 0.6 | 3.0 | Sample, Light Garlicky, treated. | Natural | 60.6 | 21 |
| Final | Sept. 23 | 13.68 | 55.3 | .4 | 4.9 | 4 Light, Garlicky Weevily, limed. | Lime | 50.3 | 32 |

BIN U-6 (FORCE VENTILATED. SEE TEXT FOR AIR-PATH LENGTH)

| Initial | July 17 | 15.9 | 56.3 | 0.5 | 0.4 | Sample, Light Garlicky. | Natural | 69.7 | 21 |
| Final | Sept. 23 | 12.54 | 56.6 | .5 | 4.5 | 3, Light Garlicky, Weevily. | ____do____ | 51.0 | 32 |

[1] All wheat of class soft red winter and subclass red winter.

ILLINOIS TESTS, 1938

VERTICAL FLUES

In 1938 two of the 300-bushel bins were fitted with vertical flues to test this type of natural ventilation. Although flues of this type had proved very inefficient in drying wheat in previous tests in Kansas and North Dakota, it was thought that under Illinois weather conditions, where the usefulness of ventilation was more apparent in cooling than in drying the grain, they might have some advantages. It is also usually easier to fill and empty a bin fitted with vertical than with horizontal flues. Bin U-2, with tight wood walls and floor, was fitted with three vertical galvanized flues, 8 inches in diameter, set on the east and west, or long axis of the bin, so that each flue was approximately in the center of mass of a third of the bin (fig. 14).

The flues were perforated with small holes from 6 inches above the floor up to about 12 inches below the upper wheat surface. The tops of all three flues were connected by means of 8-inch circular ducts to a common suction cowl on the bin roof. Directly opposite the vertical flues three rectangular openings, each 10 inches wide and 45 inches high, were made in the north and south walls, and covered

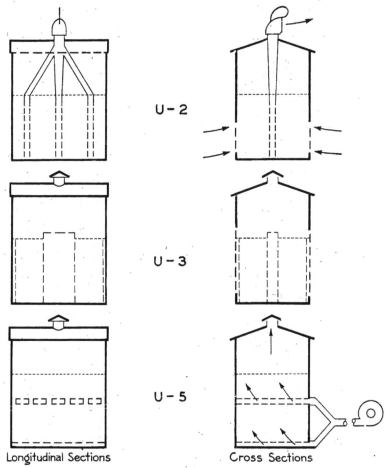

Longitudinal Sections   Cross Sections

FIGURE 14.—Diagram of ventilated bins used at Urbana, Ill., in 1938.

with fly screen backed up by hardware cloth. These wall openings provided inlets for air which could pass out through the vertical flues and suction cowl.

Another bin, U–3, was lined on the inside of the studs with hardware cloth and fly screen, instead of the matched lumber as used in the other bins. The door, however, in the north wall was of wood. In the center of the bin a vertical flue, 8 inches wide and 32 inches long, of fly screen on a wooden frame was installed, extending from the floor to several inches above the grain line. No suction or pressure cowls were used to assist the flow of air. The flue did not extend through the floor.

Bins U-2 and U-3, with an unventilated bin, U-4, for check pur- poses, were each filled on July 1, 1938, with 14.7-percent moisture content wheat from the same lot; U-2 receiving 294 bushels, U-3, 286 bushels, and U-4, 296 bushels.

The mean temperature of the wheat in all three bins at the time of filling was about 81° F. In U-4, the unventilated bin, there was very little change in temperature until August 1, when the wheat in the north half of the bin, which originally had a moisture content about 0.25 percent above the bin average, started to increase in tem- perature until on September 12, after being in storage 73 days, a high of 105° was reached, the wheat turned sour, the bin was emptied, and the wheat disposed of. On the other hand, the wheat in bin U-2 dropped within a week to approximately 77°, and except for slight increases during short warm spells, kept this temperature until cold weather arrived. Bin U-3 did not cool so rapidly as U-2, and on August 1, when the mean temperature of the wheat was 76°, began to increase very slowly in temperature, although never exceeding a peak of 85° reached on August 12 near the bottom of the bin on the north side.

TABLE 8.—*Grade-factor data, percent germination, and fat acidity of Illinois wheat samples, 1938*

BIN U-2 (3 VERTICAL FLUES CONNECTED TO SUCTION COWL)

Sample	Date	Mois- ture	Test weight per bushel	Dock- age	Dam- age	Grade [1]	Odor	Ger- mina- tion	Fat acid- ity
		Per- cent	*Pounds*	*Per- cent*	*Per- cent*			*Per- cent*	*Units*
Initial____	July 6, 1938	14.67	59.6	0.3	([2])	2 Tough, Light Gar- licky.	Natural___	85.5	16.1
Latest____	Jan. 5, 1939	14.55	59.3	.4	0.7	____do_____	Fumigant_	80.7	21.8

BIN U-3 (8 INCHES BY 32 INCHES VERTICAL FLUE, NO COWL)

| Initial____ | July   5, 1938 | 14.7 | 59.8 | 0.3 | 0.6 | 2 Tough, Garlicky____ | Natural_ _ | 84.0 | 20.7 |
| Latest____ | Jan.   5, 1939 | 14.88 | 58.7 | _____ | 4.0 | 2 Tough, Light Gar- licky. | Fumigant_ | [3] 15.5 | 31.4 |

BIN U-4 (UNVENTILATED CHECK ON BINS U-2 AND U-3)

Initial____	July 5, 1938	14.74	59.9	0.3	_____	2 Tough, Light Gar- licky.	Natural___	86.8	19.7
Inferior [4]_	Sept. 12, 1938	14.96	57.7	.5	10.0	Sample, Tough, Light Garlicky, Weevily.	Musty____	45.5	32.5
Balance [5]_	_____do_____	14.65	59.5	_____	_____	2 Tough, Light Gar- licky, Weevily.	Natural___	77	20.4

BIN U-5 (FORCE-VENTILATED WITH NORMAL AIR)

| Initial____ | July   9, 1938 | 14.5 | 59.6 | 0.3 | ([6]) | 2 Tough Garlicky_____ | Natural___ | 81.2 | 18.6 |
| Final_____ | Sept. 30, 1938 | 13.81 | 59.8 | _____ | ([6]) | 2 Light Garlicky, Wee- vily. | Weevily___ | 67.5 | 20.2 |

[1] All wheat except that in U-5 of class soft red winter and subclass red winter. Wheat in U-5 of class mixed and subclass mixed wheat.
[2] Trace.
[3] The fumigant used may have contributed to the large drop in germination.
[4] Sample from north central part of bin, representing about 67 bushels of heating wheat.
[5] Sample from balance of bin, including no wheat from heating or inferior portion.
[6] Negligible.

As shown in table 8, there was no drying of the wheat in either U–2 or U–3, perhaps because of the unfavorable drying conditions that are characteristic of this area. Moisture determinations on samples taken from near the screened walls and the flues over a period of several months, indicated that even in those locations very little moisture was removed. The wheat was still stored in these two bins on April 6, 1939.

FORCED VENTILATION

It will be remembered that in the 1936 tests, ventilation with air introduced at the floor and forced through the entire depth of wheat, accomplished some drying, although the principal effect was to move moisture from the lower to the upper part of the wheat mass. The following year the forced-ventilation bin was modified by the addition of two layers of flues in such a way as to increase the wheat surface exposed to air and to decrease the distance of air travel. However, although the amount of moisture removed per cubic foot of air was increased slightly, the system had the disadvantage of being hard to install because of the many flues.

In 1938, to simplify the bin construction, bin U–5, with tight wood walls and perforated metal floor, was fitted with a single layer of flues with centers 40 inches above the floor. These flues, unlike those used in previous years, were each 18 inches wide and 3 inches thick, covered top and bottom with fly screen. Eighty percent of the bin horizontal cross section was exposed to air. In operation, the air was to enter the wheat both through the floor and the flues, all of it escaping through the upper wheat surface. By adjusting the pressure at the floor and flue inlets, the air entering at the floor could be made to pass up, gathering moisture as it rose, and, at the flues, before it had become saturated, be diluted and lowered in relative humidity by fresh air.

Bin U–5 was filled on July 9, with 304 bushels of mixed wheat (hard red winter 60 percent, soft red winter 40 percent), moisture content 14.5 percent, the depth of wheat being 6 feet 3 inches. On July 14 the blower was started, forcing approximately 700 cubic feet per minute through the floor and 400 cubic feet per minute through the flues, and was run continuously for a 31-hour period. Only normal-temperature air was used. A series of samples for moisture determination, taken from both above and below the flues, showed that more moisture was removed from above the flues than from below, as shown in table 9.

TABLE 9.—*Forced ventilation of bin U–5, 1938*

	Average moisture content	
	Above flues	Below flues
	Percent	*Percent*
Before starting blower July 14	14.12	14.62
After 19½ hours of ventilation	13.68	14.47
After 31 hours of ventilation	13.67	14.21

When ventilation was stopped, although the average bin moisture content had been reduced from 14.37 percent to 13.89 percent, there still were several points below the flues that had over 14 percent of moisture.

The cost of electricity for operating the blower used in removing moisture by this method was 1.6 cents per bushel per 1-percent moisture reduction; 100.2 kilowatt-hours being used at 2.5 cents per kilowatt-hour. No estimates were made of the labor cost or the investment in blower, motor, and flue system.

During the time the blower was operated the bin was held on the scales so that hourly weighings could be made. The weight loss as well as the dry-bulb temperature and relative humidity of the ventilating air are shown in figure 15. As the moisture content varied so greatly within the bin, and as part of the moisture lost was undoubtedly from the wood of the bin walls, the weight loss cannot be converted

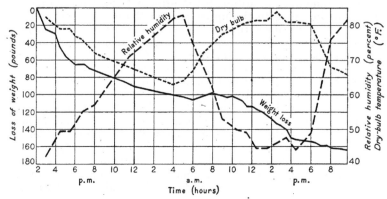

FIGURE 15.—Temperature and relative humidity of ventilating air and weight loss from bin U–5, July 14 and 15, 1938.

into a grain-drying-rate curve, expressed in percent loss per hour, but the curve does indicate very plainly a smaller loss of moisture during the night.

From the initial and final moisture content and weight of the wheat, the moisture removed during the 31-hour period was calculated to be 0.36 grain per cubic foot of air.

RESULTS OF ILLINOIS TESTS

In comparing the results of the 3 years' tests the weather conditions in the different seasons and the condition of the different lots of wheat must be considered. The summers of 1936, 1937, and 1938 were all warmer than average. July and August 1936 were exceptionally hot and dry, as indicated in table 10 by the departures of the temperatures and relative humidities from normal. Such months are favorable for drying wheat exposed to air because of the low relative humidity, but the high temperatures cause rapid deterioration of masses of wheat of high moisture content. Thus comparisons between ventilated and nonventilated bins were more than normally favorable to the ventilated bins in these months and in September 1937 and September 1938. No whole month of the three summers was as unfavorable for tests of drying as might have been expected in an ordinary season, though there were some short periods of high relative humidity.

The initial wheat moistures, germinations, and fat acidities given in tables 5, 7, and 8 show that because of higher moisture content and incipient damage, as indicated by lower germination and higher fat acidity, the wheat used in 1937 was much more difficult to store without damage than that used in 1936 or 1938; the wheat with 17-percent initial moisture stored in bins U-3 and U-8 in 1937 was more difficult to keep in good condition than that with about 16-percent moisture stored in bin U-3 in 1936. Wheat used in the other tests was only slightly moister than was safe for ordinary storage in unventilated bins.

The horizontal flues tested in bins U-3 and U-4 in 1936 and U-3 and U-8 in 1937 retarded heating but did not remove moisture rapidly enough to avoid eventual damage to wheat containing 15.5 percent or more of moisture. It is probable, however, that in most years bins with ventilators of the size and spacing used in U-3 or U-4 would have safely stored, through the fall months, sound wheat of 15-percent moisture. Means must be provided for tightly closing these flues when the bins are fumigated to destroy insects.

TABLE 10.—*Summer temperatures and relative humidities and departures from normal at Springfield, Ill., 1936–38*

Year	July				August				September			
	Mean		Departure from normal		Mean		Departure from normal		Mean		Departure from normal	
	Temperature	Relative humidity [1]	Temperature	Relative humidity [1]	Temperature	Relative humidity [1]	Temperature	Relative humidity [1]	Temperature	Relative humidity [1]	Temperature	Relative humidity [1]
	°F.	Percent	°F.	Percent	°F.	Percent	°F.	Percent	°F.	Percent	°F.	Percent
1936	86.2	48	+9.7	−17	82.5	57	+8.0	−12	72.8	74	+5.2	+2
1937	78.0	63	+1.5	−2	79.5	70	+5.0	+1	67.6	65	0	−7
1938	80.2	66	+3.7	+1	79.4	70	+4.9	+1	71.9	69	+4.3	−3

[1] Found by averaging 8 a. m. and 8 p. m. readings.

The vertical flue used in combination with ventilated walls in bin U-3 in 1938 permitted the escape of some heat, but removed little moisture. It is doubtful if this arrangement is as efficient as horizontal flues in either drying or cooling the wheat. Bin U-2, with screened wall panels and vertical flues spaced about 3 feet on centers and connected to a suction cowl removed little moisture but provided enough air movement through the wheat to prevent a rise in temperature. This arrangement with suction cowls appears to be somewhat more efficient than similarly spaced horizontal flues not connected to cowls. Ventilated walls have the disadvantage of making it difficult to close the bin tightly for effective fumigation to destroy insects.

Forcing air up from the bottom of the bin through 6 or more feet of wheat as in bins U-5 and U-6 in 1936 was of small value, for little moisture was removed from the bin and the upper layers of the wheat were made excessively damp by condensation of the moisture driven out of the lower layers. This was true with both unheated and heated air, even under unusually favorable conditions of relative humidity for Illinois.

Forced ventilation during the warmest parts of the day, with the air path through the wheat not exceeding 3 feet, as in bins U-5 and U-6 in 1937, was effective in drying and cooling wheat of 16-percent moisture content before much damage occurred, though some difficulty was found in getting even drying in the different parts of the bins. These were the most successful tests of the series. However, good drying was also obtained by the method of introducing fresh air at two levels as in bin U–5 in 1938.

The Illinois tests were of special interest because they provided a means of checking weight losses against moisture losses. In each case the weight loss was more than the calculated moisture loss, but no further study was made of these losses.

STORAGE OF HARD RED WINTER WHEAT [6]

Most of the hard red winter wheat grown in Kansas is harvested by combines and in some years presents a storage problem because of excessive moisture and green and immature kernels. Hays, in the west-central section of the State, is in the center of the combine territory, and because of the facilities of the Fort Hays branch station, was selected as the location for the wheat-storage investigations in this section. The studies reported by the Kansas Agricultural Experiment Station,[7] were made at this station in 1929, 1930, and 1931. Several of the bins and other equipment used were still available. In addition to this equipment, the station each year loaned several thousand bushels of wheat for experimental use for 3 to 6 months, and the technical staff advised and assisted in this work.

KANSAS TESTS, 1936

Three 1,000-bushel, round, steel bins were used in 1936 to check various methods of ventilation. These bins were all 14 feet in diameter and 8 feet high to the plates, and had conical steel roofs (fig. 16). Bin H–5, with tight walls had a floor of fly screen laid on an angle-iron grid so that air could enter the wheat at the bottom and pass up through the grain and out through a ventilator in the bin roof. This ventilator was not of the suction type but was merely a protected opening. The additional cost of a ventilated floor over that of a new tight floor need not exceed 3 cents per bushel.

Bin H–6 (fig. 17) the same size and shape as H–5, had a tight, metal floor and perforated metal walls, and was fitted with a vertical galvanized flue 20 inches in diameter in the bin center, connected at the upper end to a suction-type cowl kept pointed away from the wind by a vane. The bottom part of the flue was perforated to within a short distance of the upper wheat surface with a number of small holes, spaced 1¼ inches on centers and approximately one-sixth of an inch in diameter, with a total area of about 0.2 percent of the flue area. The ventilating air could enter the wheat through the perforated walls, pass through approximately 6 feet of grain, and pass out through the suction cowl by way of the vertical flue, or through some portion of the perforated wall.

---

[6] In the planning and conduct of the experiment, valuable assistance was given by L. C. Aicher, superintendent of the Fort Hays branch, F. C. Fenton of the Agricultural Engineering Section, C. O. Swanson of the Milling Industry Section, and other members of the Kansas Agricultural Experiment Station. The U. S. Department of Agriculture representatives were A. D. Edgar, associate agricultural engineer, and W. R. Swanson, junior agricultural engineer. These men and their assistants were largely responsible for the conduct of the field work. Analysis for grade of the samples taken were made in the Kansas City, Mo., office of the Grain Division, Bureau of Agricultural Economics.

[7] SWANSON, C. O., and FENTON, F. C. THE QUALITY OF WHEAT AS AFFECTED BY FARM STORAGE. Kans. Agr. Expt. Sta., Tech. Bull. 33, 70 pp. illus. 1932.

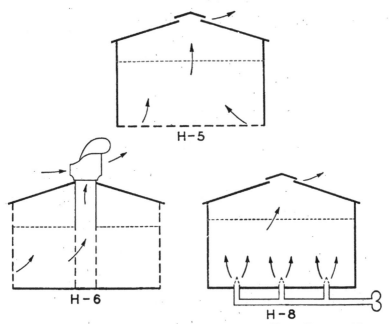

FIGURE 16.—Diagram of ventilated bins used at Hays, Kans., 1936.

FIGURE 17.—Bin H–6, with perforated walls and suction cowl. Bins H–5, H–7, and H–8 were similar except for the perforated walls and suction cowl.

159482°—40——5

The only forced-ventilation system tested was in bin H–8, a 1,000-bushel, round metal bin with tight walls and floor. Air could be forced into the wheat through five 1¼-inch sand-point-tipped pipes extending 1 inch above the floor—one in the center and four equally spaced 4 feet 3 inches about the center. The blower used was of the positive displacement gear type, driven by a ½-horsepower electric motor.

Bin H–7, exactly the same size and shape as the three just described, but having no ventilation, was used as a check bin.

Grain was delivered directly to the experimental bins from combines operating in the vicinity. While attempts were made to harvest grain that had between 14 and 15 percent of moisture, the moisture content of the various truckloads varied greatly, some being as low as 10 and some as high as 18 percent. The higher moisture content grain had many green and immature kernels. Each truckload was divided about equally among the four bins (H–5, 6, 7, and 8) by means of a divider so that, although all bins had approximately the same kind and moisture content of wheat, the wheat in each bin varied greatly within itself, having layers of damp and dry grain.

The four bins were filled through the divider on June 25 and 26, and the first average samples were taken within a day or two. No samples were taken at this time to locate the wet layers in the bins. In table 11 is shown the number of bushels stored, depth of grain, and initial temperature in the center of each bin.

TABLE 11.—*Amount stored, days in storage, depth of wheat in bin and initial wheat temperature, Kansas, 1936 tests*

Bin No.—	Quantity stored	Time in storage [1]	Depth of wheat		Initial temperature at bin center
	Bushels	*Days*	*Ft.*	*In.*	*°F.*
H–7	780	12	6	4	111
H–5	782	[2] 741	6	0	105
H–6	739	10	5	9	113
H–8	780	95	6	4	115

[1] Days in storage before first abnormal odor appeared.
[2] Bin H–5 did not go out of condition, and no abnormal odor appeared.

The center temperature in bin H–7 increased steadily from 111° to 127° F. in 10 days, then decreased to 122° on July 8, when, after 12 days in storage, an area in the center amounting to 25 to 30 bushels was found to be slightly sour and out of condition. The grade data for this sample, as well as for the balance of the bin, are given in table 12. The sample from the heating area had a moisture content well above the bin average, and samples taken 5 days later substantiated this difference.

The wheat in bin H–5, with the screen floor, from the same lot of wheat as in H–7, decreased in 2 weeks from a temperature of 105° in the center to 77°. The moisture also dropped about 1 percent in the same period, and there was probably additional benefit derived from transfer of moisture from wet layers to dry layers, resulting in a more even moisture content throughout the bin. This transfer evidently did not take place in the unventilated bin H–7. The wheat in bin H–5 was left in storage until November 25, when it was moved in order to weigh it, and then returned to the same bin where it remained until July 7, 1938. At this time, still in good condition (table 12), it was removed to another bin. Numerous samples taken during the storage period indicated that the wheat was always in good condition.

TABLE 12.—*Grade-factor data, percent germination, and acid value of Kansas 1936 wheat samples*

H-7 (UNVENTILATED CHECK BIN)

Sample	Date	Moisture	Test weight per bushel	Dockage	Damage	Grade[1]	Odor	Germination	Fat acidity[2]
		Pct.	*Lbs.*	*Pct.*	*Pct.*			*Pct.*	*Units*
Initial	June 27	12.8	55.5	1.1	2.0	4	Natural	86	8.84
Inferior[3]	July 8	14.4	52.4	1.2	2.0	Sample, Tough	Slightly sour		
Balance[4]	...do	12.8	55.5			4	Natural		
Inferior[5]	July 13	15.2	50.5	3.2	15.0	Sample, Tough	Slightly sour	2.0	28.29
Balance[6]	...do	12.8	55.7			4	Natural	64.0	13.3

H-5 (TIGHT WALLS, FLOOR OF SCREEN)

Sample	Date	Moisture	Test weight per bushel	Dockage	Damage	Grade[1]	Odor	Germination	Fat acidity[2]
Initial	June 27	12.9	56.0	1.14	1.8	3	Natural	85	9.08
Last	July 7[7]	11.31	56.1	.5	.0	3	...do	63.46	20.21

H-6 (TIGHT FLOOR, PERFORATED WALLS, CENTRAL PERFORATED FLUE, AND SUCTION COWL)

Sample	Date	Moisture	Test weight per bushel	Dockage	Damage	Grade[1]	Odor	Germination	Fat acidity[2]
Initial	June 27	13.0	55.3	1.3	1.6	4	Natural	80	9.03
Inferior[8]	July 6	13.3	51.0	1.8	7.0	Sample	Sour		
Inferior[9]	July 10	14.6	53.3	1.7	4.0	Sample, Tough	Slighty sour		
Inferior[10]	July 13	15.4	48.5	4.0	10.0	...do	Musty	11.0	17.96
Balance[11]	...do	11.8	55.4			4		82.0	11.4

H-8 (FORCED VENTILATED THROUGH 5 FLOOR OPENINGS)

Sample	Date	Moisture	Test weight per bushel	Dockage	Damage	Grade[1]	Odor	Germination	Fat acidity[2]
Initial	June 27	13.1	55.7	1.0	1.8	4	Natural	90	8.08
Average[12]	Sept. 29	11.6	55.7	.6	2.0	Sample	Sour	88	13.3
Do.[13]	Nov. 25	11.7	55.7	1.2	1.9	...do	...do		

[1] All wheat of class hard red winter, and subclass dark hard winter.
[2] Fat acidity, calculated from acid value determination and average oil content of hard red winter wheat.
[3] First sample showing sour odor, from area of about 25 to 30 bushels in center of bin having temperature 122° F.
[4] Balance of bin (excluding inferior portion).
[5] Sample of inferior portion (about 100 to 125 bushels in center of bin), taken at time bin was emptied. Temperature of bin center 105° F.
[6] This sample included no wheat from inferior part of bin.
[7] 1938.
[8] Sample from 10 to 15 bushels "heating" portion (temperature 102° F.), near central ventilator flue, about halfway up.
[9] Sample from 20 to 30 bushels in same location.
[10] Sample from 75 to 100 bushels in same location but spreading toward walls. Bin temperature down to 90° F. in warmest spot.
[11] This sample included no wheat from inferior part of bin.
[12] First sample having sour odor.
[13] Sample taken when bin was emptied.

Bin H-6, with perforated walls and central flue connected to a suction cowl, was filled with wheat averaging 13-percent moisture, but varying greatly in the bin, and having an initial high wheat temperature in the center, a few inches from the flue, of 113° F. This point, always the warmest in the bin, dropped to 102° by July 6, after 10 days in storage, but at that time the wheat developed a sour odor in the same area, amounting to 10 to 15 bushels. The spoiled and sour area increased rapidly in size as shown by the condition of successive samples (table 12). When the bin was emptied the wheat next to the central flue and in the wet layers was somewhat caked (fig. 18). It

FIGURE 18.—Interior of bin H–6, showing bin partly emptied. Note thermocouple wires used for reading temperature of the wheat.

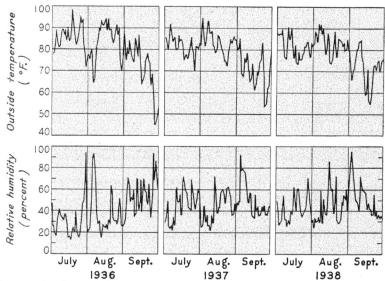

FIGURE 19.—Average daily temperature and relative humidities at Hays, Kans.,
for July, August, and September, 1936, 1937, and 1938

should be noted that the heating wheat had a moisture content higher
than the bin average.

The force-ventilated bin, H–8, filled with wheat averaging 13.1-
percent moisture content, but also varying greatly from one level to
the next, was filled on June 26 with 780 bushels of wheat from the same
source as that used in the other bins. The blower was started immediately, forcing air through the wheat at the rate of 90 cubic feet
per minute. The blower was operated continuously for 10 days, in

the 240 hours forcing through the grain approximately 1,296,000 cubic feet of air. The average temperature of the air was 81° F., and the average relative humidity 44 percent. The average moisture content of the wheat was lowered in the 10 days, from 13.1 percent to 12.2 percent, representing a loss of 448 pounds of water. By calculation, each cubic foot of air removed 2.4 grains of moisture. As only average samples were taken it is not known if the wheat dried uniformly or how much the high-moisture levels were reduced by transferring moisture to the drier areas.

The highest temperature in the bin, in the center 1 foot below the upper wheat surface, was lowered from 115° F. at the time of filling to 77° in only 6 days. Despite the low temperatures and comparatively low average moisture content, the wheat developed a sour odor after 95 days in storage, and a sampling 2 months later confirmed this condition. A baking test made on a sample drawn September 29 indicated a stale odor in flour, dough, and bread.

Average samples were taken from all the bins in 1936 when they were filled and at intervals thereafter, in order to watch changes in the condition of the stored wheat as indicated by its grade factors, germination, and fat acidity. These data are shown in table 12.

Figure 19 shows the average daily temperatures and relative humidities at Hays.

KANSAS TESTS, 1937

The fact that the grain used varied so greatly in moisture content and that the air temperatures were higher than the average for Hays, made the conditions under which a ventilated bin had to operate in 1936 very adverse. However, it should be remembered that the same conditions might occur whenever the combine is used and the grain is stored without mixing. The same groups of bins were used in the 1937 studies, but more emphasis was placed on forced and wind-assisted natural ventilation than in 1936. Several of the bins were divided by partitions through the center, on the north and south axis, to separate two different types of ventilating systems used in the same bin (fig. 20).

TESTS WITH NATURAL VENTILATION

As in the previous year, a plain, unventilated bin was used as a check on the performance of the ventilating systems. This check bin H–1E, was the east half of a round, 10-foot-diameter metal bin.

Three new bins, H–10, H–11, and H–12, were built each with a floor area 4 by 5 feet and double wooden walls, lined on the inside with aluminum foil to retard vapor movement and cooling, and more nearly simulate a full-size bin. Each had a floor of perforated metal, similar to that of bin H–5 in 1936. The three bins were all built together in a line, H–10 and H–12 being on the ends (fig. 22). No ventilating cowls were used. Circulation of air over the top of the grain was provided through openings under the eaves. The roof sloped from the top of a 10-foot wall on the east to a 6-foot wall on the west. This set of bins was used to test the efficiency of a perforated floor in keeping wheat at various depths, and for comparison with the performance of bin H–5 the previous year.

Bin H–8W, the west half of a round, 14-foot-diameter, tight-wall-and-floor, metal bin, was fitted with flues and a suction cowl in an effort to take advantage of the comparatively high wind movement in the locality. Four 3-inch diameter perforated metal flues were set vertically on a line parallel to the partition and 3 feet from it. The flues were spaced 2 feet 4 inches apart in the line. The upper ends of these 4 flues opened to the outside air through a common duct or manifold. Five similar flues were placed 2 feet 6 inches apart in a line only 1 foot from the partition and 2 feet east of the first row, and similarly, 3 other flues (2 feet 6 inches between centers) were placed in a line parallel to the first row and 2 feet 2 inches west of it. The upper ends of the latter 8 flues were connected to a common suction-type cowl by means of a manifold. Thus 12 vertical flues were arranged so that air, passing into the center row of flues, would

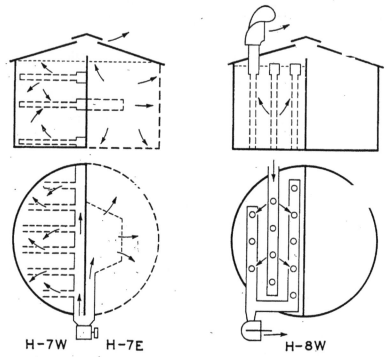

H-7W   H-7E   H-8W

FIGURE 20.—Diagrams of ventilated bins at Hays, Kans., in 1937

pass horizontally through about 2 feet 6 inches of wheat, to the flues connected to the suction cowl.

The 1936 bin H-6, with tight floor, perforated walls, and perforated central flue, was divided in the center by a north and south partition, making two half-round 500-bushel bins. The partition passed through the center of the vertical flue. The west half of this structure, H-6W, was used for the 1937 studies. A 16-inch propeller-type fan, operated by a 1/12-horsepower motor, was placed just below the suction cowl, with its axis in the vertical plane, to assist the wind in drawing air through the wheat.

Wheat from nearby farms was not used in 1937, but grain was shipped by rail from Enid, Okla., arriving in Hays June 29. This method of handling resulted in the wheat being well mixed and of a much more even and somewhat higher moisture content than in 1936. The initial temperature when stored was also lower. The bins were filled immediately through a divider in order to get the same kind of wheat in each bin. In table 13 is given the number of bushels stored, depth of grain, and initial temperature in the center of each bin. The moisture content was 15.4 percent.

TABLE 13.—*Amount stored, days in storage, depth of wheat, and initial wheat temperature in Kansas experimental bins, 1937*

Bin No.—	Quantity stored	Time in storage [1]	Depth of wheat	Initial temperature at bin center	Bin No.—	Quantity stored	Time in storage [1]	Depth of wheat	Initial temperature at bin center
	Bushels	*Days*	*Ft. In.*	*° F.*		*Bushels*	*Days*	*Ft. In.*	*° F.*
H–1E	230	14	6 8	85	H–8W	500	56	7 10	84
H–10	64	16	4 0	87	H–6W	500	74	8 0	85
H–11	96	20	6 0	87	H–7E	500	[2] 58	8 0	82
H–12	128	19	8 0	87	H–7W	500	[2] 59	8 0	84

[1] Days in storage before first abnormal odor appeared.
[2] These bins did not go out of condition but were disposed of Aug. 27 when they were dried.

As was expected, the wheat in the unventilated check bin H–1E increased in temperature very rapidly near the center of the bin, going from 85° F. when filled to 117° on July 15. The first musty odor was noticed in this bin July 13, after 14 days in storage. As the wheat was so well mixed, no pockets or layers of out-of-condition wheat were found. Samples taken from the upper, lower, and central levels of the bin showed that most of the drop in the average moisture content of the bin was due to loss from the upper part. When moved on July 16 much of the wheat was caked throughout the bin.

The temperature of the wheat in the centers of the series of bins H–10, 11 and 12, at the time of filling, was 87° F. These temperatures rose in H–10 to 109° in 16 days, in H–11 to 121° in 20 days, and in H–12 to 119° in 19 days, after which each bin was emptied because the wheat had developed a musty odor. It was so badly caked that some parts had to be broken up with a shovel before it could be removed.

The temperature of the wheat in bin H–8W, having the wind-assisted ventilating system utilizing vertical flues, increased very slowly at the center from 84° to 94° F. on August 9, after which it cooled slowly to 85° after being in storage 56 days. The first abnormal odor was also noticed at this time. The bin was not emptied until September 11 when some badly spoiled wheat, having a moisture content of 17.1 percent was found around the bottoms of the intake flues. There was also a small amount of caked wheat on the walls of the bin. The average wind velocity during the period June 30 to August 9 was 6.7 miles per hour.

Bin H–6W, with perforated walls and 20-inch-diameter central ventilating flue connected to fan-assisted suction cowl, was somewhat more successful in keeping the stored grain than was bin H–8W. The fan was operated continuously, except for a period of a few days when the fan motor was being overhauled. A point in the center of the mass of wheat, initially having a temperature of 85° F., dropped within a few days to 78°. However, a point a few inches from the central flue reached a temperature of 101° on August 6. The average moisture content was reduced from 15.3 percent to 13.5 percent, a loss of 1.8 percent, before the wheat developed a musty odor, compared to a reduction from 15.5 to 14.1, a 1.4 percent loss, in bin H–8W for the period prior to going out of condition. When the bin was emptied on September 11, 2½ months after filling, a layer of rotten wheat about 1 inch deep was found over the entire floor of the bin, and the wheat was also caked slightly in the center of the bin.

FORCED-VENTILATION TESTS

For tests of forced-ventilation systems, a perforated-wall, tight-bottom metal bin, 14 feet in diameter, was divided by a partition to form two half-round bins of the same size (fig. 20.) Bin H–7E, the east half, was fitted with a perforated floor and an air-pressure chamber in the center of the bin, made with wood framework covered with perforated metal. The chamber was 1 foot high or thick, and in plan was a half hexagon 3 feet 6 inches in the narrow dimension. The air path through the wheat was about 3 feet 6 inches to the side, top, or bottom of the stored grain. The air chamber was connected through a 10-inch pipe to a blower for forced ventilation.

The walls of the west half of the bin (H–7W) were lined with heavy building paper to close the perforations. Three layers of flues (fig. 20) each made of two 1- by 4-inch boards, set on edge about 4 inches apart, and covered top and bottom with perforated metal, were placed in this bin, the bottom layer 9 inches, the center layer 4 feet, and the top layer 6½ feet above the floor. The distance between centers of flues in each layer was 2 feet. The bottom and top flues were connected through a manifold to a blower for forced ventilation. The center layer passed through the wall to provide an outlet for air forced into the bottom and top layers. Air could also escape through the upper wheat surface.

Bins H–7E and H–7W were filled through the divider at the same time as were the naturally ventilated bins previously discussed. For the amount of wheat, depth, initial temperature, and days in storage, see table 13, and for the grade data, germination percentage, and fat acidity of the wheat at the beginning and end of storage in each bin, see table 14.

The ventilating systems of both forced-ventilation bins were supplied with air at the same time from the same blower at the rate of 540 cubic feet per minute to each bin. The blower was operated continuously from 9 a. m. June 30 until 7 p. m. July 7, and then intermittently until July 11. The hours of ventilation and the condition of the ventilating air are given for these two bins in table 15, up to and including July 6, at which time the bin was sampled for moisture determinations (see table 14).

In the 159 hours of ventilation, each bin received approximately 5,151,600 cubic feet of air at the rate of about 540 cubic feet per minute. In H–7E the wheat moisture content of the 500 bushels stored was reduced from 15.45 percent to 11.0 percent, a drop of 4.45 percent. Each cubic foot of air removed a calculated 2.1 grains of moisture from the wheat. The wheat in bin H–7W was reduced from 15.25 percent to 11.0 percent, a drop of 4.25 percent, in the same 159-hour period. In this case each cubic foot of air removed 1.97 grains of moisture. Samples made from various levels in the bin indicated that the moisture was removed more evenly from H–7W, equipped with horizontal flues, than from the bin with the central chamber. However, the dampest wheat found in H–7E had only

TABLE 14.—*Grade-factor data, percent germination, and fat acidity of Kansas wheat samples, 1937*

H-1E (UNVENTILATED CHECK BIN)

Sample	Date	Moisture	Test weight per bushel	Dockage	Damage	Grade [1]	Odor	Germination	Fat acidity
		Percent	*Pounds*	*Percent*	*Percent*			*Percent*	*Units*
Initial_____	July 1	15.35	61.1	0.3	([2])	1, Tough_____	Natural__	71.0	20
Final_____	July 13	15.0	59.4	_____	4.0	Sample, Tough__	Slightly musty.	45.0	24

H-10 (BOTTOM VENTILATION, 4 FEET WHEAT DEPTH)

Sample	Date	Moisture	Test weight per bushel	Dockage	Damage	Grade [1]	Odor	Germination	Fat acidity
Initial_____	July 2	15.25	61.1	0.2	0.2	1, Tough_____	Natural__	70.2	20
Average_____	July 8	15.1	60.3	_____	.2	____do_____	___do____	55.0	22
Do_____	July 13	15.0	59.4	_____	.3	2, Tough_____	___do____	38.2	24
Final_____	July 16	15.5	58.2	_____	5.0	Sample, Tough__	Musty___	16.5	33

H-11 (BOTTOM VENTILATION 6 FEET WHEAT DEPTH)

Sample	Date	Moisture	Test weight per bushel	Dockage	Damage	Grade [1]	Odor	Germination	Fat acidity
Initial_____	July 2	15.75	61.2	0.2	([2])	Sample [3]_____	Natural__	71.2	21
Average_____	July 7	15.4	60.6	_____	.4	1, Tough_____	___do____	60.2	22
Do_____	July 13	15.0	59.3	_____	.3	2, Tough_____	___do____	32.0	22
Final_____	July 20	14.9	59.0	_____	.1	Sample, Tough__	Musty___	25.0	32

H-12 (BOTTOM VENTILATION, 8 FEET WHEAT DEPTH)

Sample	Date	Moisture	Test weight per bushel	Dockage	Damage	Grade [1]	Odor	Germination	Fat acidity
Initial_____	July 2	15.2	61.6	0.1	([2])	1, Tough_____	Natural__	66.7	21
Average_____	July 7	15.3	60.8	_____	.2	____do_____	___do____	64.5	21
Do_____	July 13	15.4	59.6	_____	.5	2, Tough_____	___do____	30.8	24
Final_____	July 19	14.8	59.4	_____	4.2	Sample, Tough__	Musty___	26.0	26

H-8W (NATURAL VENTILATION, VERTICAL FLUES, WIND ASSISTED)

Sample	Date	Moisture	Test weight per bushel	Dockage	Damage	Grade [1]	Odor	Germination	Fat acidity
Initial_____	June 30	15.5	61.3	0.15	([4])	1, Tough_____	Natural__	69.0	20
Final_____	Aug. 24	14.1	59.5	_____	5.0	Sample, Tough, Weevily.	Musty___	29.7	30

H-6W (ONE CENTRAL VERTICAL FLUE, 20 INCHES IN DIAMETER, WITH SUCTION COWL AND SMALL FAN)

Sample	Date	Moisture	Test weight per bushel	Dockage	Damage	Grade [1]	Odor	Germination	Fat acidity
Initial_____	June 30	15.3	61.0	0.14	([4])	1, Tough_____	Natural__	67.0	20
Average_____	Sept. 7	13.5	59.5	_____	5.0	3_____	___do.[5]__	30.2	33
Final (inferior)____	Sept. 11	14.7	59.2	_____	5.4	Sample_____	Musty___	2.2	49
Final (balance)__	___do_____	12.8	59.3	.2	4.2	3_____	Natural__	48.2	27

H-7E (FORCE VENTILATED, CENTRAL HALF HEXAGON AIR CHAMBER)

Sample	Date	Moisture	Test weight per bushel	Dockage	Damage	Grade [1]	Odor	Germination	Fat acidity
Initial_____	June 30	15.45	61.4	0.13	([4])	1, Tough_____	Natural__	70.7	20
Average_____	July 6	11.0	60.7	_____	.1	1_____	___do____	67.5	22
Final (average)___	Aug. 27	10.5	61.4	.2	1.5	1_____	___do____	72.2	21

H-7W (FORCE VENTILATED, HORIZONTAL FLUES)

Sample	Date	Moisture	Test weight per bushel	Dockage	Damage	Grade [1]	Odor	Germination	Fat acidity
Initial_____	June 30	15.25	61.4	0.13	([4])	1, Tough_____	Natural__	72.0	19
Average_____	July 6	11.0	61.2	_____	.5	1_____	___do____	69.5	20
Final_____	Aug. 27	([6])	61.8	.2	1.0	1_____	___do____	71.5	21

[1] All wheat class hard red winter and subclass dark hard winter.
[2] None.
[3] Sample grade because of moisture content over 15.5 percent.
[4] Trace.
[5] The sampler stated that this sample had a slightly musty odor, but this was not confirmed by the Federal grain supervisor.
[6] No moisture determination for this sample, but the moisture content of an average sample taken Aug. 24, 1937 showed 9.7 percent moisture.

13.0 percent of moisture after ventilation, and it is probable that the greater ease of building and installing the central chamber in the granary and of filling and emptying the bin would offset any advantages obtained from more even drying. When the two bins were emptied on August 25 and 27, the wheat was dry and hard, and in excellent condition. The temperatures of the wheat near the center of each bin, about 84° F. when filled, dropped after 22 hours of ventilation to 61°. The cooling due to evaporation in the first few days of ventilation sometimes caused a lowering of the exhaust-air temperature by as much as 40° below the inlet air.

TABLE 15.—*Hours of operation and conditions of ventilating air in forced-ventilation bins H-7E and H-7W, Kansas, 1937*

Date 1937	Opera-tion	Air					
		Temperature			Relative humidity		
		Average	Maximum	Minimum	Average	Maximum	Minimum
	Hours	*°F.*	*°F.*	*°F.*	*Percent*	*Percent*	*Percent*
June 30	15	78	98	53	29	59	10
July 1	24	85	101	71	30	36	20
July 2	24	85	104	66	45	64	20
July 3	24	83	97	66	52	95	18
July 4	24	88	106	68	32	70	8
July 5	24	92	108	76	27	42	72
July 6	24	90	106	72	30	50	14
Total	159						

KANSAS TESTS, 1938

To provide better drainage and to give the ventilated bins more exposure to the wind, the experimental bins were moved to the top of a hill southeast of the Fort Hays station buildings, for the 1938 studies.

TESTS WITH NATURAL VENTILATION

It was decided to repeat in 1938 the 1936 test of screen-bottom ventilation, using the original bin H-5, which had tight metal walls and a galvanized screen bottom, but to attempt to get well-mixed wheat of approximately the same average moisture content as placed in this bin in 1936. To further test and if possible increase the efficiency of this type of ventilation, bins H-10, H-11, and H-12 were equipped with suction-type ventilating cowls on the roofs to assist the flow of air up through the wheat (figs. 21 and 22). The ventilating spaces around the eaves were closed.

The wheat stored in the naturally ventilated bins in 1938 was purchased from a local elevator. The wheat placed in H-5 was well mixed, but that in H-10, H-11, and H-12 varied somewhat in moisture content. All bins were filled on June 30 and July 1, and the first samples were drawn a few days later. The bin numbers, bushels stored, depth of wheat, days in storage, and initial wheat temperature are given in table 16. The grade, germination, and acid-value determinations for the initial and successive samples from all the experimental bins are given in table 17.

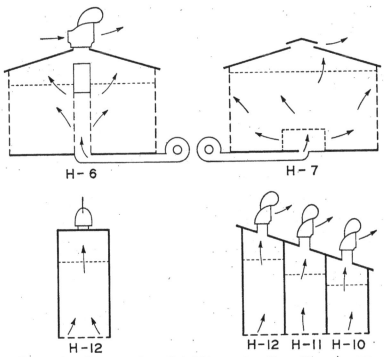

H-6   H-7

H-12   H-12 H-11 H-10

FIGURE 21.—Diagram of ventilated bins used at Hays, Kans., in 1938.

FIGURE 22.—Bins H–10, H–11, and H–12, showing suction-type ventilating cowls. Note thermocouple lead wires beside door 11.

TABLE 16.—*Bushels stored, days in storage, depth of wheat, and initial wheat temperature for Kansas experimental bins, 1938*

Bin No.	Quantity stored	Time in storage [1]	Depth of wheat	Initial temperature at bin center
	Bushels	*Days*	*Ft. In.*	*°F.*
H-5	992	191	8 0	86
H-10	62	190	3 8	88
H-11	94	190	5 8	84
H-12	125	190	7 0	83
H-6	825	192	6 9	85
H-7	934	135	7 5	86

[1] Bin H-7 was emptied after being in storage 135 days. All other bins were still in storage on Jan. 7, 1939 and figures in this column represent time from filling to that date.

TABLE 17.—*Grade-factor data, percent germination, and fat acidity of Kansas wheat samples, 1938*

H-5 (BOTTOM VENTILATION, 8-FOOT DEPTH WHEAT)

Sample type	Date	Moisture	Test weight	Dockage	Damage	Grade [1]	Odor	Germination	Fat acidity
		Percent	*Pounds per bushel*	*Percent*	*Percent*			*Percent*	*Units*
Initial	July 5, 1938	13.0	56.8	1.0	(²)	3	Natural	[3] 87.2	[3] 11.5
Intermediate [4]	Aug. 1, 1938	12.8	56.7	.5	(²)	3	do	[3] 86.0	[3] 18.3
Latest	Jan. 7, 1939	12.3	56.8	.4	(²)	3	do		

H-10 (BOTTOM VENTILATION, SUCTION COWL, 3 FEET 8 INCHES WHEAT DEPTH)

Initial	July 5, 1938	14.4	57.5	0.5	(²)	3 Tough	Natural	[5] 84.9	[3] 11.5
Intermediate [4]	Aug. 1, 1938	13.3	57.7	.5	(²)	3	do	[5] 90.5	[5] 18.6
Latest	Jan. 7, 1939	12.4	57.7	.5	(²)	3	do		

H-11 (BOTTOM VENTILATION, SUCTION COWL, 5 FEET 8 INCHES WHEAT DEPTH)

Initial	July 5, 1938	14.68	57.5	0.7	(²)	3 Tough	Natural	[5] 90.4	[3] 11.5
Intermediate [4]	Aug. 1, 1938	13.0	57.7	.5	(²)	3	do	[5] 86.4	[5] 18.6
Latest	Jan. 7, 1939	12.7	57.2	.7	(²)	3	do		

H-12 (BOTTOM VENTILATION, SUCTION COWL, 7 FOOT WHEAT DEPTH)

Initial	July 5, 1938	14.5	57.5	0.6	(²)	3 Tough	Natural	[5] 86.8	[5] 12.78
Intermediate [4]	Aug. 1, 1938	12.7	57.7	.5	(²)	3	do	[5] 86.1	[5] 18.0
Latest	Jan. 7, 1939	12.9	57.8	.5	(²)	3	do		

H-6 (FORCE-VENTILATED, CENTRAL VERTICAL FLUE)

Initial	June 30, 1938	16.2	54.1	0.8	(²)	Sample [6]	Natural	[3] 84.2	[3] 11.3
Intermediate [4]	Aug. 1, 1938	12.4	56.0	.9	(²)	3	do	[3] 83.9	[3] 15.7
Latest	Jan. 7, 1939	10.6	55.6	1.3		4	do		

H-7 (FORCE-VENTILATED, AIR CHAMBER ON FLOOR)

Initial	July 1, 1938	13.75	57.0	0.4	(²)	3	Natural	[7] 83.1	[7] 11.6
Intermediate [4]	Aug. 1, 1938	10.7	56.7	.4	(²)	3	do	[7] 84.1	[7] 15.3
Latest	Nov. 12, 1938		56.9	.6	0.6	3	do	87.5	15.9

[1] All wheat of class hard red winter and subclass dark hard winter.
[2] None.
[3] No germination and fat acidity determinations were made on the average sample. These figures are the average of 13 special samples drawn on the same date from various parts of the bin.
[4] Average samples taken during storage. Samples from bins H-6 and H-7 were taken immediately after blower was stopped.
[5] Same as [3] except only 5 special samples were averaged.
[6] Sample grade because of high moisture content.
[7] Same as [3] except 12 special samples were averaged.

Bin H–5, filled with wheat of 13.0 percent average moisture content, cooled from an initial center temperature of 86° to 80° F. in 1 week; and at no time thereafter did any point in the wheat, except near the metal walls, exceed this temperature. Drying was very slow, as shown in table 17, averaging about 0.25 percent per month during July and August, but the wheat remained in good condition.

The series of bins, H–10, H–11, and H–12, were filled with wheat of somewhat higher moisture content, averaging about 14.5 percent. The initial center temperature of 83° in H–10 dropped to 80° in 2 weeks and never exceeded this level. H–12 took 3 weeks to cool off from 83° to 80°, and H–11 4 weeks to lower from 84° to 80°. None of these bins showed any indications of heating during the storage period. During the first month of storage H–10 lowered in average moisture content 1.1 percent; H–11, 1.7 percent; and H–12, 1.8 percent; the most drying taking place in the bin with the greatest depth of wheat.

Although the four perforated-bottom bins, H–5, H–10, H–11, and H–12, did not all have the same moisture content or depth of wheat, it is interesting to note the variation in moisture reduction in the different levels, and in the average of the bin during the same period, as shown in table 18. In each case there was much more drying at the top and bottom of the wheat than within the mass.

TABLE 18.—*Moisture loss from various levels in bottom-ventilated bins, 1938*

Bin. No.	Level in bin	Moisture in wheat sampled—		Moisture loss
		July 5	Sept. 1	
		Percent	*Percent*	*Percent*
H–5	Average	13.0	12.5	0.5
	6 feet above floor	12.95	12.2	.75
	4 feet above floor	13.25	13.0	.25
	6 inches above floor	13.3	12.5	.8
H–10	Average	14.4	12.8	1.6
	3 inches below upper surface	14.8	12.8	2.0
	2 feet above floor	14.4	13.7	.7
	4 inches above floor	15.1	12.2	2.9
H–11	Average	14.7	13.0	1.7
	6 inches below upper surface	15.0	12.2	2.8
	3 feet above floor	14.6	14.0	.6
	4 inches above floor	15.0	12.2	2.8
H–12	Average	14.5	12.9	1.6
	1 foot below upper surface	13.8	12.9	.9
	3 feet 4 inches above floor	14.9	14.2	.7
	4 inches above floor	14.6	11.7	2.9

TESTS WITH FORCED VENTILATION

The quick drying in the forced-ventilation bins of 1937 suggested further work along this line, with special attention to be paid to simplifying the systems and utilizing blowers to be found on the ordinary mechanized farm, such as blower elevators and silo fillers.

The partition was removed from 1937 bin H–6W, leaving it as it was in 1936, with a 20-inch-diameter perforated flue set vertically in the center of a 14-foot-diameter perforated-wall metal bin. The perforations in the flue were increased in diameter so that they totaled about 4 percent of the flue surface. The upper end of the flue was blocked about 1 foot below the upper wheat surface and the bottom connected through a 6-inch-diameter pipe to a grain blower driven by a 3-horsepower motor, larger than necessary, but the only one available. Air forced into the central flue had to pass out into the wheat and through the perforated bin walls and upper wheat surface. Another perforated-wall, tight-bottom, 1,000-bushel metal bin, H-7, was prepared for forced ventilation by building in the center of the floor a square air chamber, 4 feet on the side and 2 feet high, of perforated metal supported on a wooden frame. The chamber was connected through a duct to a heavy-duty fan belted to a 1-horsepower electric motor (fig. 23). The passage of air was from the fan to the air chamber, and through the wheat to the perforated walls and upper wheat surface.

Bin H-6, with central perforated flue receiving air from the grain blower, was filled with 825 bushels of well-mixed wheat, averaging 16.2-percent moisture content, on June 30. The blower was started at once and ran continuously for 31 days, when, on August 1, the average moisture content had dropped to 12.4 percent. Samples taken at 5-day intervals showed that during the first 5 days the moisture content near the flue was reduced to 9.2 percent while the wheat at the wall and halfway between the central flue and wall

FIGURE 23.—Fan used in 1938 for forced ventilation of bin H-7.

was reduced only a fraction of a percent. At the end of 2 weeks, the wheat at a point equidistant from bin walls and central flue had lowered to 10.3-percent moisture while the wall areas did not dry out until after the blower was stopped. For the air conditions during the ventilating period see figure 19.

Approximately 1,020 cubic feet per minute of air, under a static pressure in the delivery pipe of 1.1 inches water gage, were forced through the wheat during the 31-day period. Using this figure for the amount of air furnished, and the average wheat moisture content before and after drying (table 17) gives a total of 1,940 pounds of water removed by 45,532,800 cubic feet of air. This is 0.29 grain of moisture per cubic foot of air.

Temperature within the bin dropped in the areas where drying was progressing at the most rapid rate. The wheat near the flue dropped as low as 61° F. during the first 5 days of blower operation but after the moisture content was lowered to 9.2 percent, the temperature in this region followed the air temperature. The wheat at a greater distance from the flue then cooled until it, too, had been dried after which its temperature followed the air temperature.

Uniformly high-moisture-content wheat could not be obtained for bin H-7, but rather than discontinue the test, wheat with an average initial moisture content of 13.75 percent was used. This wheat varied greatly in moisture content, ranging from 12 percent to 17 percent. The blower was started immediately after filling and ran continuously for 30 days, developing a pressure in the air chamber of 1.7 inches water gage, and forcing 2,400 cubic feet of air per minute through the wheat. As was the case with bin H-6, drying took place at the most rapid rate near the fresh-air inlet, and during this period the wheat temperature was lowered. The last wheat to be dried was near the walls and upper wheat surface.

Samples indicated that after the first 10 days of blower operation no moisture was removed. On July 11 the average bin moisture content was 10.4 percent, 0.3 percent lower than it was when the blower was stopped 20 days later. During this 10-day period, the average air temperature was 85° F. and the average relative humidity 36.5 percent. A total of 1,991 pounds of water were removed during the first 10 days by 34,560,000 cubic feet of air, an average of 0.41 grain per cubic foot. The wheat was disposed of November 12, at which time it was in excellent condition.

The bin numbers, bushels stored, depth of wheat, days in storage, and initial wheat temperature at bin center for both forced and naturally ventilated bins are given in table 16.

The average temperatures and relative humidities at Hays during July, August, and September 1938, are given in figure 19.

RESULTS OF KANSAS TESTS

In making comparisons between the performances of the various bins in each of the three storage seasons, the differences in the weather conditions and the quality of the wheat must be considered. From table 19 covering the 3 years that these tests were run, it can be seen that the temperatures for the 2 months after harvest average approximately 5° warmer than usual. The humidities during these months were more favorable in 1936 than in the following 2 years.

TABLE 19.—*Weather conditions at Hays, Kans.*[1]

Year	July				August				September			
	Mean		Departure from normal		Mean		Departure from normal		Mean		Departure from normal	
	Temperature	Relative humidity	Temperature	Relative humidity	Temperature	Relative humidity	Temperature	Relative humidity	Temperature	Relative humidity	Temperature	Relative humidity
	°F.	Percent	°F.	Percent[2]	°F.	Percent	°F.	Percent[2]	°F.	Percent	°F.	Percent[2]
1936	83.8	41	−5.4	−23	83.6	41	−5.9	−23	70.5	67	−1.1	+2
1937	82.0	52	−3.6	−12	83.8	51	−6.1	−13	71.4	57	−2.1	−8
1938	80.1	55	−2.7	−9	83.6	48	−5.9	−16	73.0	58	−3.6	−7

[1] Records from the Dodge City, Kans., Weather Bureau station.
[2] Averages for 8 a. m. and 8 p. m., 1889–1913.

The grain used in 1936 was of low quality, because it contained a large percentage of shrivelled kernels. The moisture content averaged approximately 13 percent, but some portions were much higher due

to the fact that the grain was hauled in directly from the combines.
In contrast to this, the wheat used in the following 2 years had a
moisture content ranging 1½ to 3 percent higher than in 1936, but
the grain was of good quality and was more uniformly mixed.

Bearing these differences in mind it is possible to rank the bins
approximately in the order of their performance with relation to the
weather and the condition of the grain. On this basis the results
seem to indicate that in this area wheat can be stored safely for several
years in tight unventilated bins if the moisture content is 13 percent
or less. In order to store wheat safely that has a moisture content
in the range of 13.0 to 14.5 percent, it appears necessary to provide
ventilation. This may be accomplished in many ways, but the
following types of wind-ventilated bins have proved effective in field
tests. These are arranged in order of efficiency, with the most
efficient first. (1) Bins having a system of horizontal flues with
alternate layers connected to suction and pressure cowls and so
spaced that the maximum air travel between the flues is 2 feet. (2)
Bins having a suction or pressure cowl which is connected to a perforated chamber so proportioned and located in the center of the mass
of grain that the maximum distance the air travels through the wheat
is 3½ feet. These bins must have perforated floors and preferably
perforated walls. (3) Bins having a suction cowl connected to the
space above the grain in such a fashion that the air is drawn through
a perforated floor and not more than 8 feet of grain. (4) Bins having a
suction or pressure cowl connected to a perforated central chimney so
the air enters through perforated walls and passes through not more
than 6 feet of grain. Any of the naturally ventilated types enumer-
ated above can be provided with power ventilation to prevent damage
to grain having more than 14.5 percent of moisture. If the weather
is favorable for drying wheat at 18.0-percent moisture content can be
stored in ventilated bins if the air travel is very short and the blowers
have high capacity.

VENTILATION STUDIES IN THE HARD RED SPRING WHEAT AREA [8]

The ventilation investigations in North Dakota were conducted at
Fargo, in the eastern section of the State, and in the hard red spring
wheat area. The experimental bins were located on the northwest
corner of the State Agricultural College campus. Approximately
6,000 bushels of hard red spring wheat for the 1936 and 1937 experiments were loaned by the State mill and elevator of North Dakota.
Two thousand bushels of wheat of the same class for the 1938 experiment were loaned by the North Dakota Agricultural Experiment
Station, which also cooperated in the conduct of the work.

The bins were placed in a circle with an inside radius of 25 feet,
leaving a distance between bins of approximately 15 feet. No bin
was shaded at any time during the day. They were exposed to winds
from the north, west and southwest, but somewhat sheltered from the

[8] Valuable assistance in the planning and conduct of the work was given by H F. McColly of the Department of Agricultural Engineering, T. H. Hopper and R. H. Harris of the Department of Cereal Technology.
and other members of the North Dakota Agricultural Experiment Station staff. The U. S. Department of
Agriculture representatives were C. F. Kelly, associate agricultural engineer in 1936 and 1937 and M. G.
Cropsey, junior agricultural engineer in 1938. Grade analysis of the samples taken were made in the Min-
neapolis, Minn., office of the Grain Division, Bureau of Agricultural Economics.

FIGURE 24.—General view of bins at Fargo, N. Dak., showing elevator and grain divider.

south and east. A grain divider, mounted on a 40-foot tower in the center of the circle, was arranged so that any lot of wheat could be divided equally among 6 to 12 bins (fig. 24.).

NORTH DAKOTA TESTS, 1936

Seven ventilated bins were built and tested in 1936, some of the systems being very much alike. Bins F–5 and F–6 were the east and west halves, respectively, of a 1,000-bushel round, perforated-wall, metal bin, divided in the center, on the north-south axis, by a corrugated-metal partition. Bin F–5 had a perforated floor of the type furnished by the manufacturer, with slots 1 inch long, spaced 3 inches on centers one way and 1½ inches the other way. The floor of bin F–6 was also perforated, but the area of the perforations was about 10 times greater than in F–5.

Three bins, F–7, F–9 and F–11, were equipped with horizontal flues of the type developed in North Dakota in 1929. Each flue was made of two 1- by 4-inch boards set on edge, about 5 inches apart, and covered top and bottom with fly screen. Bin F–7 was the east half of a round metal bin, with eight flues running north and south, arranged as shown in figure 25. Bin F–9, the west half of a 1,000-bushel single-wall wood bin, and F–11, the west half of a 1,000-bushel double-wall wood bin, each had eight ventilating flues of the North Dakota type installed with the same horizontal and vertical spacing as F–7.

Bin F–10, the east half of a 1,000-bushel wooden bin was equipped with a ventilating system using vertical flues. Eight flues, of 1- by 4-inch boards and fly screens, were set vertically in two rows extending in a north and south direction, four flues in each row. The lower ends of these flues rested on two horizontal flues having both ends open, and the upper ends of the vertical flues extended above

the wheat.   Air could enter the bottom horizontal flues and pass up
through the vertical flues.

In order to utilize the power available from wind pressure, bin F–8,
the west half of a 1,000-bushel, round, tight-wall, metal bin, was

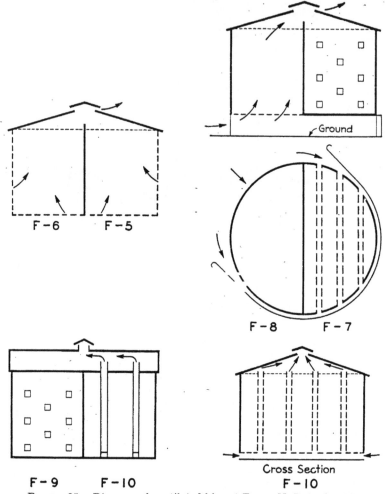

FIGURE 25.—Diagram of ventilated bins at Fargo, N. Dak., in 1936.

equipped with a perforated metal floor and large pressure chamber
underneath the bin so that winds from the northwest (the direction
of prevailing winds in that locality) would force air up through the
wheat.

In 1936 the drouth ripened the grain so quickly that it was impossible to locate naturally damp wheat.   Therefore, at the State mill
and elevator, wheat of 11.8-percent moisture was mixed and transferred while a stream of water was played on it, and the moisture
content brought up to 16 percent.   The wheat was then shipped by
rail to Fargo, a distance of approximately 90 miles.   Upon arrival,

it was in doubtful condition, having started to heat in the car. However, transfer from the car to the bins cooled it and restored its normal odor.

The ventilated bins, with an unventilated metal check bin, F-1, were filled through the grain divider during the period August 28 to 31. Each bin was sampled for grade, germination, and rancidity determinations immediately after filling and at weekly intervals thereafter. Where the wheat appeared to be going out of condition or developing an abnormal odor, additional samples were taken from the doubtful areas. Numerous thermocouples enabled the keeping of constant records of temperatures within the bin.

The bushels stored, days in storage until going out of condition, depth of wheat, initial temperature, and total percent moisture removed are given in table 20.

TABLE 20.—*Quantity, temperature, and moisture of 1936 wheat in ventilated bins at Fargo, N. Dak.*

Bin No.—	Quantity stored	Time in storage	Depth of wheat	Initial temperature at bin center	Moisture		
					Initial	Final	Loss
	Bushels	*Days*	*Feet*	*°F.*	*Percent*	*Percent*	*Percent*
F-1	493	35	8	76	15.45	15.40	0.05
F-5	493	42	8	77	14.85	14.85	.55
F-6	493	42	8	72	15.25	14.72	.53
F-8	493	49	8	75	15.48	14.98	.50
F-7	481	49	8	75	15.45	14.75	.70
F-9	525	49	8	72	15.5	14.95	.55
F-11	446	49	8	75	15.52	14.8	.72
F-10	526	42	8	76	15.68	15.15	.53

The fact that the wheat had been tempered by the addition of water makes the results of the 1936 tests of value in comparing the different types of bins only insofar as removal of moisture is concerned. Although the changes in grade, germination, and rancidity were probably not the same as would have taken place in wheat having natural moisture, these determinations are given in table 21.

All the types of ventilation except that in bin F-10, using vertical flues, were effective in preventing any increase in wheat temperature. The wheat in the unventilated check bin, F-1, increased rather steadily in temperature to 95° F. at the center on September 28, after being in storage 28 days, while that in bin F-10, with the vertical flues, increased to 85° F. in 17 days, but never went beyond this point. The wheat in the other ventilated bins followed the air temperature closely, the lag between the air and wheat being the least in bin F-8, having an air chamber below a perforated floor for wind-assisted ventilation.

There was little loss in moisture from any of the bins, on the basis of average samples. Samples taken on October 23 with a probe held vertically at distances of 6 inches and 1 foot from the north, east, and south walls of bin F-5, having perforated metal walls and floors, indicated that at that time the wheat 6 inches from the wall was 1.2 percent and 1 foot from the wall was only 0.2 percent below the bin average (15.05 percent on October 23). Samples taken horizontally at levels 6 inches, 24 inches, and 48 inches above the floors in bins F-1 (unventilated check bin), F-6 (perforated walls and floor), and F-8 (wind-assisted, perforated floor), showed the moisture contents indicated in table 22.

TABLE 21.—*Grade factors, germination, and fat-acidity determinations for 1936 North Dakota wheat samples*

(Grades are based on moisture determinations made at North Dakota State College)

BIN F-1 (UNVENTILATED CHECK BIN)

Sample	Date	Moisture	Test weight per bushel	Dockage	Damage	Grade[1]	Odor	Germination	Fat acidity[2]
		Percent	*Pounds*	*Percent*	*Percent*			*Percent*	*Units*
Initial	Aug. 31	15.45	54.7	1.8	3.5	4, Tough	Natural	69	11.5
Average	Oct. 5	15.4	53.0	1.82	4.0	Sample, Tough Weevily	Musty	39	20.6
Inferior	do	18.58	49.5	2.3	6.8	Sample, Weevily	do	4	54.8
Balance	do	15.12	53.4	1.7	3.0	Sample, Tough Weevily	Sour	41	

BIN F-5 (PERFORATED WALLS AND BOTTOM)

Sample	Date	Moisture	Test weight per bushel	Dockage	Damage	Grade	Odor	Germination	Fat acidity
Initial	Aug. 31	15.4	55.0	1.5	2.0	3, Tough	Natural	70	16.6
Average	Oct. 12	14.85	54.2	2.0		Sample, Tough Weevily	Sour	93	17.2
Sour[3]	do	17.12	51.7	2.2		Sample, Weevily	do	34	
Balance[4]	do	14.80	54.5	1.5		4, Tough	Natural	62	

BIN F-6 (PERFORATED WALLS AND BOTTOM)

Sample	Date	Moisture	Test weight per bushel	Dockage	Damage	Grade	Odor	Germination	Fat acidity
Initial	Aug. 31	15.25	55.3	1.4	5.0	3, Tough	Natural	68	17.3
Average	Oct. 12	14.72	55.0	1.7		do	do	49	21.6
Sour[3]	do	16.05	53.1	1.9		Sample, Weevily	Sour		
Balance[4]	do	14.65	55.1	1.7		3, Tough	Natural		

BIN F-8 (WIND ASSISTED)

Sample	Date	Moisture	Test weight per bushel	Dockage	Damage	Grade	Odor	Germination	Fat acidity
Initial	Aug. 31	15.48	55.0	1.9	3.4	3, Tough	Natural	76	17.7
Average	Oct. 19	14.98	54.3	1.6		4, Tough	do	46	22.7
Sour[3]	do	17.54	52.7	1.0		Sample	Sour		
Balance[4]	do	14.9	54.2	1.5		4, Tough	Natural		

BIN F-7 (HORIZONTAL FLUES)

Sample	Date	Moisture	Test weight per bushel	Dockage	Damage	Grade	Odor	Germination	Fat acidity
Average	Aug. 31	15.45	55.0	2.0	4.3	3, Tough	Natural	65	17.9
Final	Oct. 19	14.75	54.8	1.9		Sample, Tough	Sour	44	21.1

BIN F-9 (HORIZONTAL FLUES)

Sample	Date	Moisture	Test weight per bushel	Dockage	Damage	Grade	Odor	Germination	Fat acidity
Average	Aug. 31	15.5	54.5	2.0	2.3	4, Tough	Natural	70	17.1
Final	Oct. 19	14.98	54.6	1.8		Sample, Tough	Sour	71	19.3

BIN F-11 (HORIZONTAL FLUES)

Sample	Date	Moisture	Test weight per bushel	Dockage	Damage	Grade	Odor	Germination	Fat acidity
Initial	Aug. 31	15.52	55.1	1.3	3.2	3, Tough	Natural	67	11.7
Final	Oct. 19	14.78	54.7	1.7		Sample, Tough	Sour	48	20.2

BIN F-10 (VERTICAL FLUES)

Sample	Date	Moisture	Test weight per bushel	Dockage	Damage	Grade	Odor	Germination	Fat acidity
Initial	Aug. 31	15.68	54.5	2.4	2.7	4, Tough	Natural	68	18.6
Final	Oct. 12	15.15	54.0	2.4		Sample, Tough	Sour	59	18.3

[1] All wheat of class hard red spring, subclass dark northern spring.
[2] Fat acidity calculated from acid value determination and average oil content of hard red spring wheat.
[3] Sample from sour area in top center of bin, comprising one-tenth of total.
[4] Sample from balance of bin.

TABLE 22.—*Moisture content of test bins on specified dates, 1936*

Date	Location	Bin F-1	Bin F-6	Bin F-8
		Percent	Percent	Percent
Sept. 28	Bin average	15.15	14.97	15.22
	6 inches above floor	14.47	13.02	13.25
	2 feet above floor	14.95	14.3	14.57
	4 feet above floor	15.57	14.7	15.00
Oct. 20	Bin average	[1] 14.85	14.78	14.98
	6 inches above floor	[1] 14.68	12.77	12.85
	2 feet above floor	[1] 14.88	14.10	14.25
	4 feet above floor	[1] 15.02	14.47	14.92

[1] These samples taken Oct. 19, 1936.

The combination of perforated floor and wind chamber to assist the flow of air up through the wheat in bin F-8 apparently were not as effective in drying as the combination of perforated walls and floor in F-6. However, in F-6, the horizontal probe included wheat from the dry area near the perforated wall and this wheat would lower the apparent moisture content of the average sample from the various levels. Both ventilating systems were effective in removing moisture from the wheat in the bottom halves of the bin, when compared with the unventilated bin, F-1.

All of the bins were emptied during the middle part of November 1936. At this time the wheat in F-1, the check bin, was caked badly through the entire depth. The wheat in bin F-10, having the vertical flues, was caked above the 2½-foot level in the mass of wheat and also directly against the flues. The bottom 2½ feet of the bin showed no caking and the wheat was not discolored. There was no caking in the bins with horizontal flues (F-7, F-9, and F-11), or in those with perforated walls and floors (bins F-5 and F-6), or in F-8 which had the chamber for utilizing the wind pressure. The average daily temperatures and relative humidities as given by the Weather Bureau Station at Moorehead, Minn., directly across the river from Fargo, are shown in figure 26. The prevailing wind during September was from the south, the average velocity being 8.9 miles per hour. In October, the prevailing winds were from the north, the average velocity being 9.4 miles per hour.

NORTH DAKOTA TESTS, 1937

As previously mentioned it was concluded from the 1936 investigations that for effective drying it would be necessary to use the wind pressure more efficiently and to decrease the distance of air travel through the wheat. Because the horizontal flues had been effective in preventing heating, if not in drying, they were tested again at Fargo.

Bin F-8, with wind chamber below a perforated floor, was left the same as in 1936, as were bins F-7 and F-11, with horizontal flues. Bin F-5, with perforated walls and floor, was remodeled for forced ventilation (fig. 27) by installing in the center of the mass of grain a half-round air-pressure chamber with a vertical height of 1 foot and a radius of 3½ feet. A 10-inch galvanized pipe extended horizontally from the air drum to the bin wall, to conduct air from a 16-inch propeller-type fan, with ½-horsepower motor. Air forced into the air drum by the fan passed through 3½ feet of wheat before leaving the bin at the walls, floor, or upper wheat surface. An air drum of

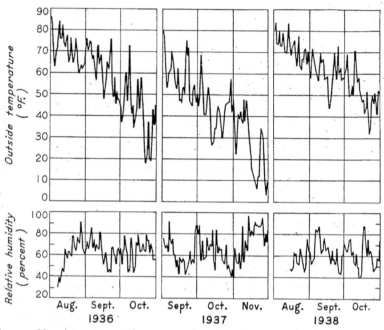

FIGURE 26.—Average outside-air temperature and relative humidity at Moorhead, Minn., during the early storage period, 1936–37–38.

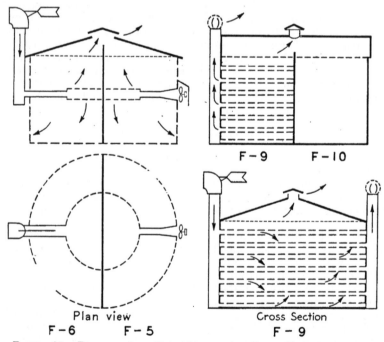

FIGURE 27.—Diagram of ventilated bins used at Fargo, N. Dak., in 1937.

the same type was installed in bin F-6, also a perforated wall and bottom metal bin, but instead of using a small fan for air pressure, a large cowl, kept headed into the wind by a vane, was placed above the bin for utilizing wind pressure (fig. 28). The cost of material for the central air chamber should not exceed 2 cents per bushel, the perforated floor 3 cents per bushel, and cowl and piping 2 cents per bushel of bin capacity.

For testing a method of wind-assisted ventilation giving a distance of air travel through the wheat less than that in bin F-6, bin F-9 was used. This was the west half of a 1,000-bushel single-wall wooden structure. Seven layers of flues were installed, each covering more than 90 percent of the horizontal cross-sectional area of the bin. One layer of flues was at the floor level and the other layers were spaced vertically at 14 inches between centers. This gave a distance in the clear, between flues, of approximately 11 inches. The first, third, fifth, and seventh layers from the floor were connected through a common manifold to a pressure cowl above the bin. The remaining three layers were connected to a turbine-type suction ventilator, also placed above the bin. The air path was from the pressure cowl to the inlet flues, through 11 inches of wheat to the outlet flues, and then to the outside via the turbine-suction ventilator.

FIGURE 28.—Bin F-6 used at Fargo, N. Dak., in 1937 showing pressure cowl. The inlet pipe is 12 inches in diameter.

Because weather conditions during harvest in 1937 were more nearly normal than in 1936, it was possible to obtain naturally moist wheat for use in the experiment. The six ventilated bins were filled on September 2 through the grain divider from two carloads of wheat averaging 15.5-percent moisture content. The initial grade, germination, and fat-acidity data from the wheat in each bin at the time they were filled, as well as from samples taken later, are given in table 24. At the time of filling on September 2, a mixture of lime and flowers of sulfur was added to bin F-7 at the rate of 0.1 pound per bushel of wheat. Oats with 10.7-percent moisture content were mixed with the wheat in bin F-8 in the proportion of 189 bushels of oats to 350 bushels of wheat. The oats absorbed enough moisture in a few hours to bring the wheat moisture content down to 14.9 percent. After 3 weeks' storage, the oats were cleaned out (September 27) with a disk mill and the wheat returned to the bin for observation.

The number of bushels stored, days in storage, depth of wheat, and initial temperature at bin center, are given for each bin in table 23.

TABLE 23.—*Quantity stored, days in storage, depth, and initial temperature of North Dakota wheat, 1937.*

Bin No.—	Quantity stored	Time in storage [1]	Depth of wheat	Initial temperature at bin center	Bin No.—	Quantity stored	Time in storage [1]	Depth of wheat	Initial temperature at bin center
	Bushels	*Days*	*Feet*	*° F*		*Bushels*	*Days*	*Feet*	*° F*
F-7	482	201	8	78	F-5	477	201	8	77
F-11	500	548	8	75	F-6	477	547	8	83
F-8	350	182	[2] 5½	[2] 62	F-9	400	201	8	79

[1] As none of the wheat went out of condition or acquired an abnormal odor, these figures represent the total time in storage before the wheat was disposed of, or until the latest sampling shown in table 24.
[2] Depth and temperature after oats were removed.

The average daily air temperature at Fargo, after the bins were filled, never exceeded 70° F., except for 2 days. This low temperature, in connection with the ventilating systems, was effective in lowering

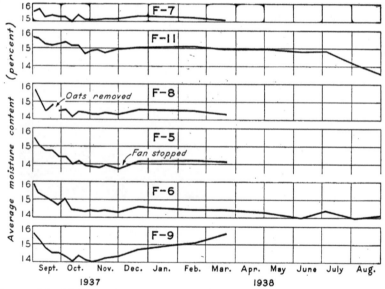

FIGURE 29.—Moisture changes in ventilated bins at Fargo, N. Dak., Sept. 1937 to Aug. 1938.

the wheat temperature and in removing any heat from the respiration of the damp wheat. The wheat in the flue-ventilated bins, F-7 and F-11, and in the bins with central drums and perforated floors and walls, F-5 and F-6, lowered in average temperature during the first 20 days in storage, at the rate of 0.7° per day. During the same period, the mixture of wheat and oats in F-8, with perforated floor and air chamber below it, lowered in temperature at the rate of 0.4° per day. The wheat in bin F-9, with the series of horizontal flues connected to pressure and suction cowls, decreased in average temperature at the rate of 0.9° per day.

Samples were taken for moisture determinations from each of the bins at weekly intervals until November 16, and at longer intervals thereafter. These determinations were used for drawing the curves of figure 29, showing the moisture changes in each bin until it was emptied or until August 31, 1938, if not emptied previously. Moisture was removed from all the bins at varying rates until about November 1, 1937. After this date, no drying was observed, the moisture content even increasing in several of the bins until, in the case of bins F–11 and F–6, which were not disposed of during the winter, drying again started in May 1938.

TABLE 24.—*Grade factor data, percent germination, and fat acidity of North Dakota 1937 wheat samples*

BIN F-7 (METAL, HORIZONTAL FLUES, WHEAT MIXED WITH LIME-SULFUR MIXTURE)

Sample	Date	Mois-ture	Test weight per bushel	Dock-age	Dam-age	Grade [1]	Odor	Germi-nation	Fat acidity
		Percent	Pounds	Percent	Percent			Percent	Units
Initial	Sept. 2, 1937	15.55	50.7	5.1	0.5	5, Tough	Natural	85.2	17
Average	Oct. 5, 1937	15.25	51.0	4.9	.4	do	do	83.7	19
Last	Mar. 21, 1938	15.05	51.8	4.9	.1	do	do	81.0	18

BIN F-11 (WOOD, HORIZONTAL FLUES, NO ADMIXTURE)

Sample	Date	Mois-ture	Test weight per bushel	Dock-age	Dam-age	Grade [1]	Odor	Germi-nation	Fat acidity
Initial	Sept. 2, 1937	15.6	53.2	3.4	(2)	4, Tough	Natural	90.2	18
Average	Oct. 5, 1937	15.35	53.2	4.2	.2	do	do	83.2	20
Do	Mar. 21, 1938	14.95	53.7	3.8	.1	do	do	76.8	17
Latest	Mar. 3, 1939	13.65	53.5	3.5	.6	4	do	66.5	26

BIN F-8 (METAL, PERFORATED FLOOR, WHEAT MIXED WITH OATS)

Sample	Date	Mois-ture	Test weight per bushel	Dock-age	Dam-age	Grade [1]	Odor	Germi-nation	Fat acidity
Initial [3]	Sept. 3, 1937	14.95	47.6		(2)	Mixed grain.	Natural	89.2	18
Average	Sept. 21, 1937	14.85	48.5		.2	do	do	85.7	21
Do	Sept. 28, 1937	14.5	53.0	5.0	.2	4	do	80.3	22
Last	Mar. 21, 1938	14.3	53.5	4.7	.6	4	do	85.0	

BIN F-5 (PERFORATED WALLS AND FLOOR CENTRAL DRUM AND 16-INCH PROPELLER FAN)

Sample	Date	Mois-ture	Test weight per bushel	Dock-age	Dam-age	Grade [1]	Odor	Germi-nation	Fat acidity
Initial	Sept. 2, 1937	15.45	53.3	4.7	(2)	4, Tough	Natural	87.2	18
Average	Oct. 5, 1937	14.4	53.2	4.1	.2	4	do	84.7	19
Last	Mar. 21, 1938	14.1	53.6	4.2	.3	4	do	83.2	

BIN F-6 (SAME CONSTRUCTION AS BIN F-5, EXCEPT USING COWL FOR AIR PRESSURE IN PLACE OF FAN)

Sample	Date	Mois-ture	Test weight per bushel	Dock-age	Dam-age	Grade [1]	Odor	Germi-nation	Fat acidity
Initial	Sept. 2, 1937	15.85	53.1	3.6	0.0	4, Tough	Natural	88.2	18
Average	Oct. 5, 1937	15.05	54.0	3.1	(2)	do	do	84.7	19
Do	Mar. 21, 1938	14.4	53.3	3.6	.3	4	do	81.2	17
Latest	Mar. 2, 1939	14.1	53.1	3.3	.5	4	do	73.5	23.8

BIN F-9 (WOOD, NATURALLY VENTILATED WITH SEVEN LAYERS OF FLUES; WIND ASSISTED)

Sample	Date	Mois-ture	Test weight per bushel	Dock-age	Dam-age	Grade [1]	Odor	Germi-nation	Fat acidity
Initial	Sept. 2, 1937	15.55	53.7	3.0	0.1	4, Tough	Natural	86.2	18
Average	Oct. 5, 1937	14.3	53.8	3.5	.3	4	do	84.5	18
Last	Mar. 21, 1938	15.6	52.2	3.1	(2)	5, Tough	do	78.7	16

[1] All wheat of class hard red spring and subclass dark northern spring except Sept. 3, 1937, and Sept. 21, 1937, samples from bin F-8, which were of class and subclass mixed grain.
[2] Trace.
[3] Moisture of wheat before adding oats 15.65 percent. Sample taken on Sept. 21 before removal of oats. Sample on Sept. 28, taken after removal of oats.

The wheat in bins F–7 and F–11, having horizontal flues, dried at a slower rate, in the fall of 1937, than did that in any of the other bins. The fact that the wheat in bin F–7 had a mixture of lime and sulfur partly filling some of the voids among the kernels probably accounts for the slower rate of drying in F–7 than in F–11. Bin F–7 was emptied on March 21, 1938, at which time the wheat was in good condition. There was no caking in any part of the bin. Bin F–11 was held for further observation.

The first moisture determination shown in the curve of figure 30 for bin F–8 was for the wheat before the dry oats were mixed with it. The later samples were tested for moisture on the basis of the cleaned wheat, so that most of the first losses in moisture should be attributed to the absorbent action of the oats and not to the ventilating system.

The fan on bin F–5 was operated continuously until noon of December 6. The wheat was removed from this bin March 21, 1938, at which time it was still in good condition. The wheat in bin F–6 was still in storage on April 1, 1939, the pressure cowl being in place during the entire period. However, snow stopped up the vertical duct leading from the cowl to the drum during part of the winter of 1938–39.

The wheat in bin F–9, having the seven layers of horizontal flues connected to suction and pressure cowls, showed about the same decrease in moisture content in the fall of 1937 as did the wheat in bins F–5 and F–6. However, as shown by the curve the average moisture content of the wheat began to rise in November.

At the time the bin was emptied March 21, 1938, the wheat near the bottom reached a moisture content of 16.8 percent, compared to 15.6 percent for the average of the bin. Part of this moisture increase may have been due to water or snow entering through the manifolds and flues. Although the moisture content of the wheat was high in the bottom part of bin F–9, there was no evidence of caking or spoiling in any part of the bin when it was emptied.

Grade-factor data, germination, and fat acidity at beginning and end of the 1937 tests are given in table 24.

NORTH DAKOTA TESTS, 1938

As stated in the preceding section, two ventilated bins, F–6 and F–11, were held over into the 1938 storage season for further observation, and the results were described under the 1937 tests.

Bin F–7, the east half of a round, 1,000-bushel commercial, metal granary, and equipped with the same horizontal flues as described under the 1937 tests, was used without alteration. The air drum and fan were removed from bin F–5, leaving it as a perforated-wall-and -floor metal bin with no other means of ventilation.

The air drum from bin F–5 was placed in the center of the 1937 bin F–8, at a distance of 4 feet above the floor (fig. 30). Air was supplied to the drum through an 8-inch metal pipe from the blower of a silo filler operated by a 3-horsepower electric motor. As the floor of this bin was perforated but the walls solid, the path of air forced into the drum from the blower was through 3½ feet of wheat to the perforated floor or upper wheat surface.

The flues were removed from the 1937 bin F–9, and the flooring removed from between the two center joists except for a distance of

18 inches in from the north and south walls.   The floor joists were 14 inches apart in the clear and extended in a north and south direction. The bottom of the two joists from which the flooring had been removed was boxed in with tongue-and-groove boards, and the open space in the floor covered with perforated metal.   The rectangular bin floor, approximately 7 by 12 feet, with a total area of 84 square feet, thus had in its center an air chamber with a surface exposed to the wheat 9 feet by 1 foot 2 inches, or a total area of 10½ square feet. Air was supplied to the air chamber through a 7-inch-diameter pipe by a farm-type pneumatic grain elevator driven by a 3-horsepower electric motor.   The path of the air forced into the air chamber, when the bin was filled with wheat, was from the perforated metal surface in the floor, upward through 8 feet of wheat, and out through the upper wheat surface.

The wheat was obtained from combines operating near Fargo and hauled directly to the bins.   This was the same procedure as was

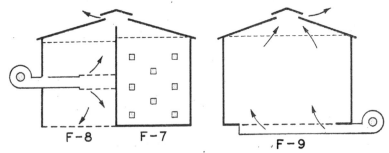

F-8   F-7

F-9

FIGURE 30.—Diagrams of ventilated bins used at Fargo, N. Dak., in 1938.

followed at Hays, Kans., in 1936 and for filling part of the bins in 1938.   Bins F-7, F-8, and F-9 were filled simultaneously through the divider from truckloads of wheat ranging in moisture content from 14.3 percent to 18 percent, the average being 16.6 percent.   The average of each bin after filling was somewhat lower.   Bin F-5, with perforated walls and floor, was filled through the divider at the same time as bin F-3, an unventilated metal bin used as a check.   The truckloads of wheat ranged in moisture content from 14.65 percent to 16.35 percent.

The bushels stored, days in storage, depth of wheat and initial temperature at bin center are given for each bin, including bin F-3, the check bin, in table 25.

TABLE 25.—*Days in storage, bushels stored, depth of wheat, and initial temperature at bin center of wheat in North Dakota, 1938, experimental bins*

Bin No.	Quantity stored	Time in storage	Depth of wheat	Initial temperature at bin center	Bin No.	Quantity stored	Time in storage	Depth of wheat	Initial temperature at bin center
	Bushels	*Days*	*Feet*	*° F.*		*Bushels*	*Days*	*Feet*	*° F.*
F-3_____	450	20	7	81	F-8_____	495	[1] 212	7	81
F-5_____	470	26	7½	81	F-9_____	505	[1] 212	7½	81
F-7 _____	505	28	8	81					

[1] Wheat still in storage on Mar. 3, 1939.

Average samples were taken at the time of filling the bins and at intervals thereafter, for grade, germination, and fat-acidity determinations. These data are shown in table 26.

TABLE 26.—*Grade-factor data, percent germination, and fat acidity of North Dakota, 1938, wheat samples* [1]

BIN F-3 (UNVENTILATED CHECK ON BIN F-5)

Sample	Date	Moisture	Test weight per bushel	Dockage	Damage	Grade [2]	Odor	Germination	Fat acidity
		Percent	Pounds	Percent	Percent			Percent	Units
Initial____	Aug. 3, 1938	14.9	56	0.9	0.0	3, Tough, Smutty__	Smutty_	81.2	24.2
Inferior [3]___	Aug. 23, 1938	15.85	54.6	.5	8.5	Sample, Smutty___	Musty__	17.5	44.9
Balance [4]_	____do_____	14.8	57.2	.5	.4	2, Tough _____	Natural_	89.0	21.8

BIN F-5 (METAL, PERFORATED WALL AND BOTTOM)

Initial____	Aug. 3, 1938	14.9	56.2	0.9	0.2	3, Tough, Smutty__	Smutty_	71.2	23.4
Inferior [5]___	Aug. 29, 1938	16.1	53.0	1.8	4.8	Sample, Smutty___	Musty_	21.7	48.3
Balance [4]__	____do_____	14.0	56.5	1.2	.7	3, Light Smutty ___	Natural_	78.0	23.1
Latest[6]____	Mar. 2, 1939	14.5	55.7	1.2	1.4	____do_____	___do_____	60.7	34.9

BIN F-7 (METAL, WITH 8 HORIZONTAL FLUES)

Initial_____	Aug. 3, 1938	15.4	56.6	1.1	(7)	3, Tough _____ Light Smutty.	Natural_	77.5	20.1
Inferior [8]___	Aug. 31, 1938	19.88	49.8	5.6	54.4	Sample_____	Musty__	4.0	57.6
Balance [4]_	____do_____	13.55	56.0	1.3	.2	3, Weevily _____	Fumigant.	87.7	18.1

BIN F-8 (PERFORATED FLOOR, FORCE VENTILATED THROUGH CENTRAL DRUM)

Initial_____	Aug. 3, 1938	15.1	56.5	0.8	0.0	3, Tough Light Smutty.	Smutty_	75.7	20.2
Average___	Sept. 3, 1938	13.45	56.8	.8	1.2	3, Light Smutty ____	___do_____	73.0	22.8
Latest_____	Mar. 3, 1939	13.45	56.6	.6	.8	____do_____	Natural_	74.7	25.0

BIN F-9 (FORCE VENTILATED THROUGH FLOOR)

Initial_____	Aug. 3, 1938	15.75	56.8	0.7	0.0	3, Tough Light Smutty.	Smutty_	74.2	24.2
Average___	Sept. 3, 1938	14.75	55.2	1.1	4.6	____do_____	___do_____	57.5	36.4
Do_____	Sept. 30, 1938	14.55	56.0	1.0	4.4	____do_____	___do_____	60.0	34.6
Latest_____	Mar. 3, 1939	14.95	55.6	.8	4.6	____do_____	_____	57.7	33.9

[1] The data from samples taken from bins F-6 and F-11, filled with wheat held over from 1937, are given in table 24.
[2] All wheat of class hard red spring and subclass dark northern spring.
[3] Sample from one-fifth of bin, upper levels.
[4] Sample from balance of bin, excluding inferior portions.
[5] Sample from one-tenth of bin, upper levels.
[6] Average sample taken from bin on Mar. 2, 1939. Wheat was moved for cooling and mixing Aug. 29, 1938, and returned to same bin.
[7] Trace.
[8] Sample from one-fifth of bin, north central location.

The wheat in the unventilated check bin F-3 began to increase in temperature, at a point near the center and 1 foot below the upper wheat surface, immediately after being placed in storage. When the temperature at this point reached 113° F. on August 23, after 20 days in storage, the bin was emptied. The grain in the upper one-fifth of the bin was very musty and had a moisture content when sampled of 15.85 percent, which was well above the average in the bin.

The grain in bin F-5 (perforated walls and floor), from the same lot as that in bin F-3, also began to rise in temperature near the top center of the bin soon after being placed in storage. When this temperature reached 105°, after 26 days in storage, the wheat was moved for cooling and returned to the same bin for further observation. When moved, the wheat in about one-fifth of the bin near the top center was found to be musty and had a moisture content of 16.1 percent. However, after thoroughly mixing with the balance of the wheat, the odor of the musty portion was not strong enough to be -detected. Very little drying occurred in any part of the bin. The wheat was still in storage on April 1, 1939.

The wheat temperatures at the centers of bins F–7, F–8 and F–9 were approximately the same in all bins when filled. In bin F–7, having eight horizontal flues, the wheat near the center of the bin began to increase rapidly in temperature immediately after filling, reaching 112° F. on August 9, after only 6 days of storage. A point nearer to the bin floor but also in the center increased more slowly to 110° after 15 days' storage. When the bin was emptied on August 31, after 28 days in storage, approximately one-fifth of the wheat in the north-central part of the bin was musty and was found to have a moisture content of 19.88 percent. A small area directly on the north wall, near the end of one of the flues, had many sprouting kernels. However, in general, the wheat appeared to be in better condition near the flues.

The blower on bin F–8 was operated for 1 hour each day during the period August 4 to 27, making a total of 24 hours of forced ventilation. The average relative humidity for the 24 days, during the period of ventilation, was 45 percent and the average air temperature 82° F. No measurements were made as to the volume of air forced through the grain, but an observation of the static pressure in the air drum showed the pressure to be 1 inch water gage.

The average wheat moisture content dropped from 15. 1 percent on August 3 to 13. 45 percent on September 3, a loss of 1. 65 percent. Samples taken from various places in the bin showed that the drying rate was greatest near the central drum and lowest at the walls, upper surface, and floor, at which points the air left the wheat. The wheat temperature throughout the bin lowered rapidly at the points where the drying rate was the highest, and none of the wheat ever showed any indication of heating. The wheat was still in storage on April 1, 1939, and in good condition.

The blower on bin F–9 was operated for several hours each day for 41 days during the period August 4 to September 28. The total number of hours of operation was 77, during which period the air relative humidity averaged 44 percent and the air temperature 79. 5° F. The static pressure in the air chamber averaged 1. 5 inches water gage. No air-volume measurements were made. The average moisture content in F–9 was lowered from 15. 75 percent on August 3 to 14. 55 percent on September 30, a loss of 1. 2 percent. Drying progressed at the most rapid rate directly over the perforated section of the floor, where the air entered, and the wheat in this part of the bin cooled rapidly. The wheat away from the inlet showed little drying, in some locations—especially near the top of the bin— actually increasing in moisture content during the first 2 weeks of ventilation. Temperatures away from the inlet locations were not lowered by the ventilating air. The areas near the top of the bin and comparatively close to the walls, not in the direct path of the air, showed a tendency to heat during the first week of storage. A temperature of 112° was reached near the east wall August 5.

RESULTS OF NORTH DAKOTA TESTS

As in other localities the weather immediately following storage is the most important factor in the safe keeping of damp wheat in ventilated bins. The average monthly temperatures, relative humidities, and departures from normal for the first 3 months of storage in 1936, 1937, and 1938, in North Dakota, are given in table 27. These records are from the Moorhead, Minn., Weather Bureau station, directly across the Red River from Fargo. It will be noted that in 1936 and 1937, the experimental wheat went into storage approximately the first of September, while in 1938, because local combined wheat was used, the storage period began almost 1 month earlier, in August. Comparing August of 1938, with September of 1936 and 1937, the relative humidity averaged 6 to 11 percent lower than in 1936 and 1937 in the first month of storage, and the temperatures averaged 10° and 13° F. higher. Because of the higher temperature, the 1938 conditions were, however, unfavorable for storage in unventilated bins. The average relative humdity during the second month of storage (October for 1936 and 1937 and September for 1938) was approximately the same in all years, but the temperature during the second month of 1938 was almost 20° higher than in 1936 and 1937. The temperatures for 1938 were several degrees above normal during August, September, and October. This was also true for September 1936. The October 1936 temperatures averaged 2.9° lower than normal. The 1937 temperatures for the first 3 storage months were almost normal.

TABLE 27.—*Weather conditions at Fargo, N. Dak.,*[1] *1936–38*

| Year | August | | | | September | | | | October | | | | November | | | |
| | Mean | | Departure from normal | | Mean | | Departure from normal | | Mean | | Departure from normal | | Mean | | Departure from normal | |
	Temperature	Relative humidity[2]	Temperature	Relative humidity[2]	Temperature	Relative humidity[2]	Temperature	Relative humidity[2]	Temperature	Relative humidity[2]	Temperature	Relative humidity[2]	Temperature	Relative humidity[2]	Temperature	Relative humidity[3]
	°F.	Pct.	°F.	Pct.	°F.	Pct.	°F.	Pct.	°F.	Pct.	°F.	Pct.	°F.	Pct.	°F.	Pct.
1936					62.2	64	+4.0	−12	41.6	63	−2.9	−15	25.2	84	−1.9	−1
1937					59.0	69	+.8	−7	44.1	68	−.4	−10	26.2	81	−.9	−4
1938	72.5	58	+6.4	−17	61.1	66	+2.9	−10	52.1	58	+7.6	−20	24.2	79	−2.9	−6

[1] Records from Moorhead, Minn., Weather Bureau station.
[2] Averages for 8 a. m. and 8 p. m., 1889–1913.

The wheat used in the North Dakota storage tests varied greatly in quality in the 3 years. The 1936 wheat, with a moisture content of approximately 15.5 percent, was of doubtful condition when placed in the bins, because of the artificial wetting of the grain and the fact that it had heated slightly in the cars before being received. The 1937 wheat averaged about the same in moisture content as the 1936 wheat (15.5 percent) but because of the low test weight (about 54 pounds per bushel) graded only 4 Dark Northern Spring. The wheat stored in one of the 1938 naturally ventilated bins averaged about the same moisture content as the 1936 and 1937 wheat. The wheat in two of the 1938 bins averaged 0.5 percent lower in moisture content and in a fourth bin 0.25 percent higher than in previous years. However, as the moisture in each of the bins varied from 14.3 percent to 18 percent, the wheat presented a storage problem greater than would well-mixed wheat with the same average moisture content. In general, all the experimental wheat, disregarding the average moisture content factor, was harder to keep in condition in storage than the general run of farm wheat—in 1936 because of the artificial wetting, in 1937 because of the low-test weight, and in 1938 because of the high-moisture wheat in pockets or layers in the bins.

The natural ventilation system using horizontal flues spaced 2 feet apart, both horizontally and vertically, tested in bins F–7, F–9, and F–11 in 1936, was of value in safely storing well-mixed and sound wheat of about 15.5-percent moisture content or about 1 percent higher than can ordinarily be stored in unventilated bins. The wind-assisted ventilating system using layers of flues spaced about 1 foot apart, as in bin F–9 in 1937, also stored 15.5-percent moisture wheat safely, but as the drying rate and dissipation of heat was so much faster than in the flue systems with wider spacing, it is probable that wheat having 16.5 percent of moisture could be safely stored in a normal year. This type of system also, because of the amount of exposure given the wheat, will, if left open, increase the moisture content of the wheat during the winter months, when both the average relative humidity and wind velocity are high; therefore, it is important that the flues be closed after cold weather sets in.

The naturally ventilated bins, using a central air chamber in connection with a perforated wall and floor bin (F–5 and F–6, 1937), did not dry or cool 15.5-percent moisture wheat as fast as did bin F–9, 1937, with the many layers of flues, but kept the wheat in very good condition. On April 1, 1939, the wheat that had been in storage in F–11 and F–6 since September 1937 was in good condition. However, the germination percentage of the F–6 wheat was 10 percent higher and the fat acidity 2 units lower than in bin F–11. It is probable that wheat somewhat higher than 15.5-percent moisture content can be successfully stored in bins of the F–6 type.

The only two forced-ventilation systems tested at Fargo were in bins F–8 and F–9 in 1938. Bin F–8, similar to F–6 except having tight walls and forcing air under 1-inch static pressure from a silage cutter and blower through 3½ feet of wheat, was effective in drying 15.1-percent-moisture wheat down to 13.45 percent after only 24 hours of blower operation, and would doubtless have dried wheat of considerably higher moisture content in ordinary weather. The periods of ventilation were about 1 hour each day during the period of lowest relative humidity. The forced ventilation system in bin F–9, utilizing a grain blower and forcing the air through 8 feet of wheat, was not as effective in drying 15.75-percent-moisture wheat, and did not entirely prevent heating.

SUMMARY AND CONCLUSIONS

The studies emphasize that under typical storage temperatures low moisture content is the most essential requirement for safe storage of wheat, and that the design and materials of a storage structure are of little importance except insofar as they influence the moisture content and temperature of the wheat. Farmers should exercise extreme caution in combining and threshing to see that the wheat goes into the bin in the best possible condition.

Other investigations carried on simultaneously with the tests reported in this circular indicate that sound wheat can be stored safely in unventilated bins for 1-year periods in western Kansas with 13 to 13.5 percent of moisture, in Illinois and Maryland with 13.5 to 14.0 percent of moisture, and in North Dakota with 14.0 to 14.5 percent of moisture. The higher moisture content permissible in North Dakota is due primarily to the lower temperatures.

These studies of the various methods and means of ventilating farm storages show that it is possible to store wheat with higher initial moisture content than in nonventilated bins, if sufficient air movement through wheat is provided to take advantage of local weather conditions. These conditions, especially wind velocity and relative humidity, influence the design of the ventilating system.

The western portion of the hard red winter wheat area is the most favorable region for drying wheat by bin ventilation. Monthly mean relative humidities at noon range from 43 to 56 percent during normal years, with the higher humidities in the winter, but during harvest and the early storage period relative humidities of 10 to 20 percent occur frequently. Monthly mean wind velocities range from 10 to 11 miles per hour.

In the hard red spring wheat area monthly mean relative humidities at noon range from 46 to 76 percent during normal years, the higher humidities occurring during the colder months. Daily minimum relative humidities of 25 to 30 percent are frequent when harvest and storage begin. Monthly mean wind velocities range from 8 to 10 miles per hour, the higher velocities generally occurring during the comparatively damp months of January, February, and March. In this area ventilating systems should not be left open during the winters, since wheat will absorb moisture from the damp air and driving snow.

In the western portion of the soft red winter wheat area monthly mean relative humidities at noon range from 48 to 76 percent during normal years, with the higher humidities in the winter. Daily minimum relative humidities of 45 to 50 percent occur during the harvest and storage season but not so frequently as lower humidities occur in the hard-wheat areas. Monthly mean wind velocities range from 9 to 10 miles per hour, the higher velocities generally occurring during the comparatively damp months of January, February, and March.

In the eastern portion of the soft red winter wheat area monthly mean relative humidities at noon range from 48 to 62 percent during normal years, the higher humidities occurring during the summer. Minimum relative humidities of 55 to 60 percent or higher are frequent when harvest and storage begin, and, as previously stated, little drying takes place under these conditions. Monthly mean wind velocities range from 6 to 7 miles per hour, the higher velocities generally occurring during the comparatively dry months of January, February, March, and April. Winter and spring are better drying times than summer.

Bins ventilated with horizontal flues such as were used in bins B-6 or B-8 in 1936 ordinarily are safe for storing wheat with initial moistures 1 percent higher than would be safe in nonventilated bins. The cost of material for these flues spaced 18 inches on centers horizontally and 24 inches vertically is 6 to 8 cents per bushel, of bin capacity. Since they can be made easily removable they are not as much of a nuisance as they may seem. They must be provided with hoods to shed rain and stoppers to close the flues tightly to keep out driving snow and humid air or to hold fumigation gases.

Closely spaced flues with pressure and suction cowls, such as were used in bins B–6 and F–9 in 1937 are more efficient than those discussed above and should be safe for storing wheat of 2 percent more moisture than in unventilated bins. The cost of material for such a system will range between 15 and 25 cents per bushel, being lower for the larger bins. Other types of ventilation using power will ordinarily be more economical.

Vertical flues did not prove as effective as horizontal flues with similar spacing.

Bins with perforated floors but without means for forcing air through the wheat were effective for equalizing moisture content between wet and dry layers of wheat at Hays, but removed very little moisture. They probably do not raise the safe average initial moisture content by more than 0.5 percent, and are not recommended for the soft red winter wheat area. The additional cost of a ventilated floor as compared to a new tight floor need not exceed 3 cents per bushel. The efficiency of ventilation through a perforated floor can be considerably increased by adding a suction cowl on the roof of the bin and tightly closing all openings above the wheat. In western Kansas such bins may be safe for wheat of 1½ percent higher moisture content than in nonventilated bins, but their effectiveness in regions of high humidity is limited.

In the drier areas the efficiency of a bottom-ventilated bin can be further increased to take care of wheat of higher moisture content by forcing air in either direction through the wheat by means of a grain blower, silo-filler blower, or any other heavy-duty fan that may be available. Because of the uneven drying resulting from power ventilating such a bin, ventilating should be continued until the layers of wheat farthest from the inlet have dried to a safe moisture content. For an illustration of uneven drying characteristic of this type of ventilation see U–5 in 1936 and F–9 in 1938.

For more efficient drying by power ventilation, means should be provided so that the air need not flow through the full depth of wheat. The simplest method of accomplishing this is by the use of a central pressure chamber and perforated floor, as in bin F–8, 1938, or H–7E in 1937, which in addition had perforated walls. If a blower of adequate capacity and a few hours of reasonably dry weather are available each day, as is usually the case in the western wheat areas, such a system should dry wheat of 18-percent moisture content without damage. Quicker and more even drying can be obtained by use of ducts instead of the central air chamber, as in bins H–7W and B–7 in 1937.

These means of air distribution can also be used to advantage with a wind-pressure cowl as in bin F–6, in 1937. Dampers should be provided for closing the cowl during the winter. The cost of material for the central air chamber should not exceed 2 cents per bushel, the perforated floor 3 cents per bushel, and cowl and piping 2 cents per bushel. The cost of material for flues as in B–7 and H–7–W will be slightly higher.

The value of any ventilation system is not determined entirely by the effectiveness with which it removes moisture and heat from the stored wheat. If it does not provide complete protection against the entrance of rain or snow, serious localized damage may develop. If wheat is to be stored for a year or more, fumigation will generally become necessary from time to time, especially in the soft red winter wheat area. Difficulty in sealing a bin because of structural features of the ventilating system or walls not only discourages the practice of fumigation, but makes it of uncertain effectiveness. Farm storages cannot, ordinarily, be inspected with sufficient frequency or thoroughness to detect the failure of an application of a fumigant. Particular attention should therefore be given to the possible interference of the ventilating system with sealing a bin for effective fumigation.

APPENDIX

PRESSURES AVAILABLE FROM WIND

Many attempts have been made to make use of wind pressure in increasing the rate of air flow through wheat in bins with natural-ventilation systems. The relation between wind velocity and its velocity pressure, calculated from the formula $V = 4005\sqrt{P}$ where V is the wind velocity in feet per minute and P is the velocity pressure of the wind, is shown in figure 31.

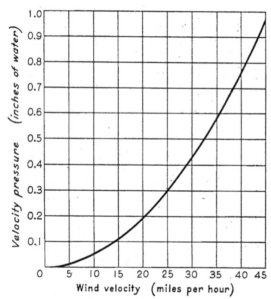

In using this chart for design purposes, it should be borne in mind that the static pressure that will be set up in a pressure chamber, into which wind is diverted by cowls and air ducts, will always be less than the velocity pressure corresponding to the wind velocity. This is partly due to the fact that the air velocity in such a chamber is not reduced to zero, but is still moving at a reduced velocity. The pressure theoretically available for conversion into static pressure is the difference between the velocity pressures of the freely moving wind and the slowly moving air in the pressure chamber. A considerable amount of the pressure theoretically available is lost by air friction in eddies formed when the wind velocity is suddenly reduced in entering the cowl. There is also a further unavoidable loss by friction between air and pipes or ducts, especially in bends or elbows where the direction of air movement is suddenly changed.

FIGURE 31.—Relation of wind velocity to pressure.

RATES OF AIR FLOW THROUGH WHEAT UNDER LOW STATIC PRESSURES

Although some data were available on rates of air flow through wheat at the comparatively high pressures attained by power blowers, used for forced ventilation, no studies had been made at the low pressures ordinarily available from the wind. In order to have some basis upon which to design naturally ventilated bins, using wind pressure, a series of experiments were run in the laboratory to determine the rate of air flow through wheat of various depths and physical conditions, at static pressures ranging from 0.025 to 0.5 inch of water, the range of pressures that would be developed by moderate to very steady winds.

The experimental apparatus used (fig. 32) consisted of a small metal bin 18 inches in diameter and 8 feet high, with a bottom of perforated metal set over an air chamber. Air was supplied to the air chamber through a 3-inch pipe from a small multivane fan. Static pressures were taken in the air chamber by means of a low-pressure air gage with a range of 0 to 0.5 inch of water, and air velocities in the 3-inch pipe with thermocouple anemometers with which velocities as low as 6 feet per minute could be read. The fan was driven by V-belts from an electric motor through a countershaft fitted with several sets of pulleys so that various pressures in the air box could be obtained by changing pulleys. In all the tests

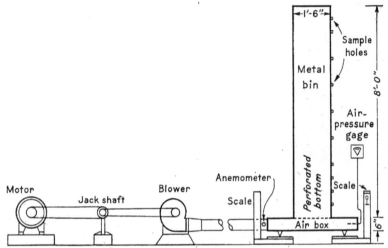

FIGURE 32.—Diagrams of laboratory bin for study of air flow and drying rates.

of air flow, No. 5 hard red spring wheat, test weight 52.7 pounds per bushel, and moisture content approximately 10 percent, was used. In filling the bin the wheat was dropped a uniform distance of 2 feet so that the grain would be packed to equal and uniform density in each test.

Various pressures were used in the air box for depths of 6 inches, 1, 1½, 2, 3, 4, 5, 6, 7, and 8 feet of wheat, and the velocity in the 3-inch pipe recorded for each depth and pressure. The amount of air supplied per square foot of bin bottom was calculated and the results shown graphically in figure 33.

Study of figures 31 and 33 shows that the quantity of air that could be forced through wheat by a comparatively brisk wind, even if it were possible to convert all the velocity pressure into static pressure would be small and suggests that in the design of naturally ventilated bins as large an area of wheat as possible be exposed to air and that the distance of air travel through the wheat be as short as possible.

EFFECT OF CHAFF ON RATE OF AIR FLOW

Farmers have been advised frequently to clean their wheat before placing it in storage in order to remove the green and wet weed seeds, and prevent the transfer of the weed-seed moisture to the wheat. Where ventilated bins are used, there is a further reason for removing dockage before storing—to increase the flow of air through the grain.

To determine the resistance offered to air flow by chaff in wheat, a quantity of dry straw was chopped in a grinder fitted with a 5⁄32-inch screen, to lengths ranging from ⅛ to 1 inch, the shorter lengths predominating. Two percent by weight of this chaff was added to the wheat in the previous experiment (which contained less than 1 percent of dockage), and a series of tests were made to measure the amount of air forced through depths of 6 inches, 1, 1½, 2 feet, and 2 feet 8 inches of the mixture by static pressures ranging from 0.025 to 0.35 inch of water. Comparison of the air-flow results shows that the rate of air flow through wheat containing 2 percent of chaff was approximately 25 percent less than through the clean wheat, under similar conditions of static pressure and grain depth.

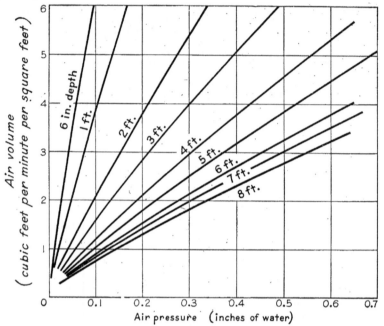

FIGURE 33.—Relation between static pressure and air flow through various depths of wheat.

EFFECT OF BIN-WALL SURFACE AREA ON RATE OF AIR FLOW

The theory has been advanced that under pressure air would have a tendency to flow up along the bin-wall surface rather than pass evenly up through the mass of wheat. To determine the effect of bin-wall area on the flow of air, the metal wall surface of the experimental bin was increased approximately 166 percent by placing three 5-inch metal pipes vertically in the bin before it was filled. Air-volume and pressure tests were run at the same pressures and wheat depths as used previously and the velocities recorded. When these results were compared with those obtained with just the bin-wall surface area, there was no difference to be noted in the amount of air forced through equal amounts of wheat under similar pressures.

EFFECT OF WHEAT DENSITY ON RATE OF AIR FLOW

Any condition that would change the ratio of voids to the amount of space taken up by kernels should have an effect on the flow of air. To determine the amount of this effect, the experimental bin was filled to the 8-foot level, the wheat being dropped a distance of 2 feet, a series of velocity readings at various pressures from 0.015 inch to 0.6 inch of water were taken, and then the wheat was well settled by pounding and shaking the bin. This lowered the wheat 3 inches, or to the 7-foot 9-inch level. The velocity readings were then repeated, using the same pressures, and the results compared. At pressures below 0.15 inch of water there was little difference between the unsettled and settled wheat. However, above 0.15 inch, even with the lower wheat level, there was a lowering of the amount of air forced through the denser or settled wheat by the same pressures of approximately 10 percent.

PRESSURE DROP IN WHEAT

When the cross sectional area of the air path through the wheat does not change, as in the case of the experimental bin used in these tests, there is a straight-line relation between air-pressure drop and lengths of air travel. This is shown in figure 34, made up from a series of tests run at seven different initial air pressures up to 1 inch of water, with 7½ feet of wheat in the bin. Holes were drilled in the

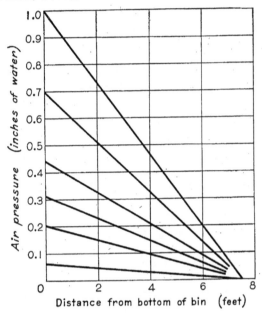

FIGURE 34.—Static air pressures at different distances above perforated bottom of experimental bin, with six different air pressures in air chamber under bin and with 7 feet 6 inches of wheat in bin.

side of the bin at the 6-inch, 1- 1½-, 2-, 3-, 4-, 5-, 6-, and 7-foot levels and the pressure taken at each of these points. However, where the cross-sectional area of the air path increases or decreases, as in a bin using flues to distribute the air, this relationship does not hold.

RATES OF MOISTURE REMOVAL

The rate of moisture removal from wheat in bins depends on the amount of air, condition of entering air, wheat temperature and moisture content, and length of air travel through wheat. In order to study the relative effect of these different factors on the drying rate, a series of tests were conducted in the laboratory, using the small experimental bin described before and illustrated in figure 32. For the drying-rate tests, the bin was mounted on scales so that the wheat could be weighed at frequent intervals to determine the weight loss. The original wheat moisture content was determined from an average sample of the entire lot of wheat under test. Soft red winter wheat was used in all cases, and to obtain the various initial moisture contents desired, the wheat was mixed with the desired amount of water and allowed to stand for at least 48 hours.

CONDITION OF AIR AND ITS EFFECT ON RATE OF MOISTURE REMOVAL

The effect of the relative humidity and temperature of the ventilating air on the drying rate, when the depth of wheat and amount of air supplied per minute per bushel of grain were constant, was determined in five tests, three using a depth of 2 feet of wheat tempered to approximately 15-percent moisture content (wet basis), and using 1.1 cubic feet of air per minute per bushel, and two tests using a depth of 2 feet 8 inches of approximately 15.75-percent-moisture content wheat, with an air flow of 0.9 cubic foot per minute per bushel. Air relative humidities and temperature were, in the case of the 2-foot depth, 82 percent and 73° F., 60 percent and 66° F., and 32 percent and 80°. Relative humidities and temperatures of 60 percent and 68°, and 30 percent and 81° were used with the 2 feet 8 inches depth of wheat. The results of these two tests are shown in figures 35 and 36. Little drying was accomplished at 60-percent relative humidity. The wheat temperature in these tests was about the same as the air temperature.

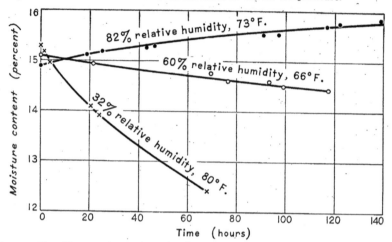

FIGURE 35.—Changes in average moisture content of wheat in a 2-foot layer, ventilated at the rate of 2 cubic feet per minute per square foot of bin bottom with air of three conditions.

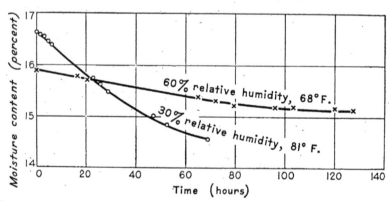

FIGURE 36.—Changes in average moisture content of wheat in a 2-foot 8-inch layer, ventilated at the rate of 2 cubic feet per minute per square foot of bin bottom with air of two conditions. Note slowing of drying rate at similar moisture contents in the thicker layer of wheat, as compared with previous test.

HEAT AND AIR MOTION IN GRAIN DRYING

Both heat and air are necessary for the rapid removal of moisture from stored grain. Moisture removal can occur only as fast as the necessary heat can be applied. The nature of the heat supply, whether natural or artificial, is not important except that a source of artificial heat makes available a greater number of British thermal units to permit drying. In order that moisture may continue to evaporate from grain, the released water vapor must be removed from the surrounding space. In still air this removal will take place by diffusion only, and therefore be very slow. When the air is moving the vapor is carried away in the air stream, so that a much faster rate of evaporation may be possible.

DRYING TESTS WITH EXPERIMENTAL BIN

In the laboratory investigations of drying rates in bins, several methods of supplying heat, in addition to that supplied by the ventilation air, were used so that the effect of heat both with and without air flow could be noted. A brief description of each of the five tests for this purpose follows.

TEST A

The 18-inch experimental bin with perforated bottom, described before, was fitted with horizontal ventilating flues at the 1-foot 8-inch and 3-feet 6-inch levels, open to the air at both ends. The flues consisted of 1- by 4-inch boards placed on 'edge about 4 inches apart, and covered on both the top and bottom with 14-mesh fly screen. Openings were cut in the bin walls to allow a free circulation of air through the flues. A 60-foot length of lead-covered hotbed cable, drawing 50 watts (equivalent to 170.76 B. t. u. per hour) was wound into three flat spiral coils. One coil was placed about 1 inch above the perforated bottom and one coil 1½ inches above each of the ventilating flues. The wheat was tempered to an average moisture content of 16.06 percent, and 370 pounds was placed in the bin. At the time the current was turned on, the wheat temperature averaged 74° F. When the test ended, after 116½ hours, the average wheat temperature had increased to 126°, with a maximum of 139° directly on the cable of the center coil. During the test the room air averaged 77° and 30 percent relative humidity. No air was forced through the bin.

TEST B

The experimental bin fitted with the same two horizontal flues as used in test A was also used in this test. However, one end of the upper flue was closed and the other end connected by a pipe to the small blower, which also supplied air to the air box beneath the bin. The air entered the wheat through the perforated bottom and the upper flue, and left by the bottom flue and the upper surface, passing through approximately 19 inches of grain. The same heating coils were used as described under test A, but were placed 4 inches below each flue and the upper wheat surface, respectively. The wheat was tempered to a moisture content of 15.16 percent and 365 pounds was placed in the bin.

At the time of filling, the average wheat temperature was 77.5°F. Electric heat (50 watts) was turned on for 21 hours, without forced-air flow. The wheat temperature at the end of that time averaged 91°. The blower was started at a speed to provide a pressure of 0.034 inch of water, and was increased later to give a pressure of 0.23 inch. These changes are shown on the drying-rate curve in figure 37. No air-velocity determinations were made. The blower was operated 21 hours, at the end of which time the average wheat temperature had lowered to 67.6°. Heating current was supplied to the cable during the entire experiment. The blower intake air averaged 76° and 22 percent relative humidity.

TEST C

In this test the cable, instead of being wound in three separate pancake-type coils, was wound around a central flue 3 inches in diameter, made up of wood and fly screen. The turns of the cable were about 1 inch apart. The flue, open at the top, extended to within 6 inches of the bin bottom. No horizontal ventilating flues were used.

The bin was filled with 367.5 pounds of wheat tempered to a moisture content of 15 percent and electric heat of 45 watts was used. At the end of 6¼ hours, very little rise in temperature had taken place so the flue was almost entirely closed at the top to prevent the escape of hot air. After another heating period of 112 hours, during which time the temperature near the coil never went higher than 115° F., the flue was filled with dry wheat and the blower started, building up a pressure of 0.04 inch of water in the air box. The heating was continued during this period also.

The drying rates under the two conditions, heat only and heat and forced air, are shown in the curve c in figure 37.

TEST D

Calcium oxide (CaO) or high calcium pulverized quicklime has a heat of hydration of 495 B. t. u. per pound, and during the hydrating process will absorb up to 25 percent of its weight of water. It thus offered possibilities as a source of

heat for evaporation, and in addition would absorb a certain amount of water
directly. An experiment was set up to determine its usefulness in grain drying.

The wheat was tempered to a moisture content of 16.95 percent and 224 pounds
were mixed thoroughly with 16 pounds of high calcium pulverized quicklime and
placed in the experimental bin equipped with perforated bottom and blower for
forced circulation. This filled the bin to the 2-foot-8-inch level. At the time
the bin was filled five thermocouples were installed, one each at the 2-, 8-, 14-,
20-, and 26-inch levels. The first reading that could be taken showed an average
wheat temperature of 78° F. In 3½ hours the average temperature had increased to 164.4°, ranging from 123.5° to over 200°. The highest temperature

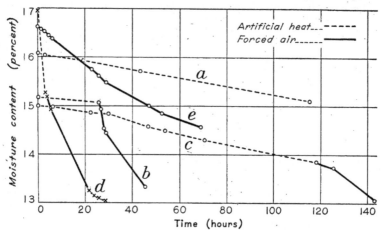

FIGURE 37.—Wheat moisture losses in experimental bin using heat and forced
air. The curves a, b, c, d, and e give results of tests A, B, C, D, and E. Dotted
lines indicate a period when artificial heat was applied without forced air.
Solid line indicates both heat and forced air or forced air only.

was near the bottom, probably because when the bin was filled it was impossible
to keep some of the lime from sifting down, causing a higher concentration at
that point. At this time (3½ hours after filling), the blower was started, delivering
1.8 cubic feet per minute per square foot of bin bottom, under a pressure of 0.42
inch of water. Reference to figure 33 shows that with a pressure of 0.42 inch
of water about 5.5 cubic feet per minute per square foot area would have been
forced through clean wheat of equal depth. The hydrated lime acted as a seal
on the voids in the wheat kernels.

The blower was operated at the same speed for 26 hours, at the end of which
time the average bin temperature had dropped to 83.6° F.

The wheat was sampled at the time of filling, when the blower was started, and
at the end of the experiment. Before the blower was started the lime had absorbed approximately 25 percent of its weight of water, or practically its ab-
sorptive capacity, which was sufficient to lower the moisture content from 16.95
percent to 15.27 percent.

TEST E

In this test, 2 cubic feet of air per minute per square foot of floor area were
forced through 2 feet 8 inches of tempered wheat in the experimental bin.
No heat was used. The air temperature averaged 81° F., and the relative humidity 30 percent.

The change in wheat moisture content in the experimental bin, used in five
different ways as to the use of heat and air, is shown in figure 37. The dotted
lines indicate a drying period where no forced circulation was used, and the
solid lines indicate that air was forced through the wheat by the blower. The
moisture loss curve for test E is shown on the chart to compare the rate of moisture
removal where forced air only is used with cases where heat without forced air
or both forced air and heat are used.

Comparison of the moisture-loss curves *a*, *b*, and *c* in figure 37 shows that before forced air was used, the drying from the application of heat alone went on at about the same rate in all three cases. As soon as forced air circulation was introduced (in *b* and *c*), the drying rate became from 4 to 10 times as great. Increasing or decreasing the amount of air affected the drying rate also.. (See test B and curve *b*, fig. 37.)

That the drying rate is increased by the addition of heat, when forced air is used, is shown by the curves of tests E, B, and D—E using no heat, B using electricity for heat, and D using the heat of hydration of lime. The amount of air per square foot of bin bottom was approximately 2 cubic feet per minute in each case. The heat approximately doubled the drying rate.

THERMAL CONDUCTIVITY AND SPECIFIC HEAT OF WHEAT

The National Bureau of Standards made several determinations of thermal conductivity and specific heat of hard red spring wheat. Under conditions of the tests there was an increase in thermal conductivity with increase in either moisture content or temperature. In table 27 the determinations are given.

TABLE 27.—*Thermal conductivity of wheat*

Mean temperature (° F.)	Moisture content	Thermal conductivity per hour per square foot per inch	Mean temperature (° F.)	Moisture content	Thermal conductivity per hour per square foot per inch	Mean temperature (° F.)	Moisture content	Thermal conductivity per hour per square foot per inch
	Percent	*B. t. u.*		*Percent*	*B. t. u.*		*Percent*	*B. t. u.*
77.7	14	0.95	79.4	23	1.04	80.5		0.88
91.2	14	.98	89.6	23	1.07	84.0		.92
87.0	12.5	.89	90.7	23	1.11	97.7		.99
97.2	12.5	.95	66.1		.84			

The specific heat determinations of wheat, as determined by the method of mixtures, are given in table 28.

TABLE 28.—*Specific heat of wheat*

Moisture content	Temperature range	Specific heat per gram per degree of temperature
	° C.	*Calories*
9.55 percent	22 to 50	0.39
21.25 percent	22 to 51	.51

ORGANIZATION OF THE UNITED STATES DEPARTMENT OF AGRICULTURE WHEN THIS PUBLICATION WAS LAST PRINTED

Secretary of Agriculture_____ HENRY A. WALLACE.
Under Secretary_____ CLAUDE R. WICHARD.
Assistant Secretary_____ GROVER B. HILL.
Director of Information_____ M. S. EISENHOWER.
Director of Extension Work_____ M. L. WILSON.
Director of Finance_____ W. A. JUMP.
Director of Personnel_____ ROY F. HENDRICKSON.
Director of Research_____ JAMES T. JARDINE.
Director of Marketing_____ MILO R. PERKINS.
Solicitor_____ MASTIN G. WHITE.
Land Use Coordinator_____ M. S. EISENHOWER.
Office of Plant and Operations_____ ARTHUR B. THATCHER, Chief.
Office of C. C. C. Activities_____ FRED W. MORRELL, Chief
Office of Experiment Stations_____ JAMES T. JARDINE, Chief.
Office of Foreign Agricultural Relations_____ LESLIE A. WHEELER, Director.
Agricultural Adjustment Administration_____ R. M. EVANS, Administrator.
Bureau of Agricultural Chemistry and Engi- HENRY G. KNIGHT, Chief.
    neering.
Bureau of Agricultural Economics_____ H. R. TOLLEY, Chief.
Agricultural Marketing Service_____ C. W. KITCHEN, Chief.
Bureau of Animal Industry_____ JOHN R. MOHLER, Chief.
Commodity Credit Corporation_____ CARL B. ROBBINS, President.
Commodity Exchange Administration_____ J. W. T. DUVEL, Chief.
Bureau of Dairy Industry_____ O. E. REED, Chief.
Bureau of Entomology and Plant Quarantine_ LEE A. STRONG, Chief.
Farm Credit Administration_____ A. G. BLACK, Governor.
Farm Security Administration_____ W. W. ALEXANDER, Administrator.
Federal Crop Insurance Corporation_____ LEROY K. SMITH, Manager.
Federal Surplus Commodities Corporation____ MILO R. PERKINS, President.
Food and Drug Administration_____ C. M. GRANGER, Acting Chief.
Forest Service_____ EARLE H. CLAPP, Acting Chief.
Bureau of Home Economics_____ LOUISE STANLEY, Chief.
Library_____ CLARIBEL R. BARNETT, Librarian.
Division of Marketing and Marketing Agree- MILO R. PERKINS, In Charge.
    ments.
Bureau of Plant Industry_____ E. C. AUCHTER, Chief.
Rural Electrification Administration_____ HARRY SLATTERY, Administrator.
Soil Conservation Service_____ H. H. BENNETT, Chief.
Weather Bureau_____ FRANCIS W. REICHELDERFER, Chief.

This circular is a contribution from

Bureau of Agricultural Chemistry and En-
    gineering_____ H. G. KNIGHT, Chief.
       Division of Farm Structures Research___ WALLACE ASHBY, Chief.